D0230219

BOURNEMOUTH
410038201

WALLACE & GROMIT™

Cracking Contraptions Manual

Illustrated by **Graham Bleathman**

Designed by **Lee Parsons**

Written by **Derek Smith**

Edited by Wallace & Gromit
using the EDIT-O-MATIC
(patent pending)

Wallace & Gromit are characters created by Nick Park.

Published in November 2010

A catalogue record for this book is available
from the British Library

ISBN 978 1 84425 958 8

Haynes Publishing, Sparkford, Yeovil, Somerset BA22 7JJ, UK
Tel: +44 (0) 1963 442030 Fax: +44 (0) 1963 440001
E-mail: sales@haynes.co.uk
Website: www.haynes.co.uk

Haynes North America, Inc.,
861 Lawrence Drive, Newbury Park,
California 91320, USA

Printed and bound in the USA

Contents

Contents

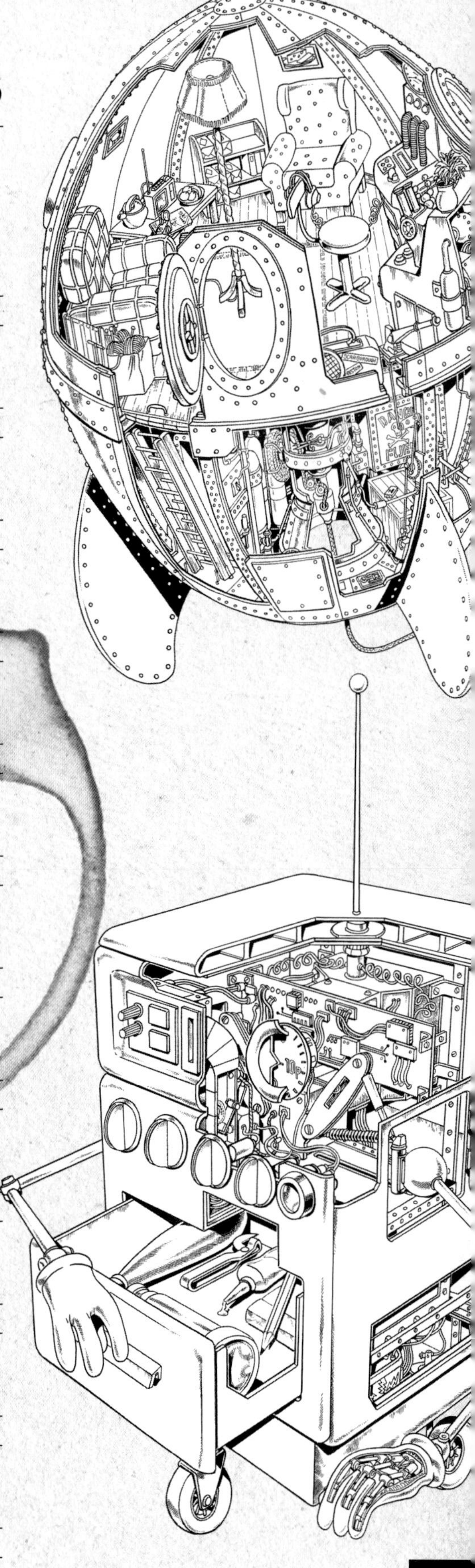

Wallace & Gromit

62 West Wallaby St,
Up North,
England

The Editor
Haynes Publishing Ltd
Sparkford
Somerset
BA22 7JJ

Dear Editor,

I can't tell you how delighted I was to receive the proofs for your new book in this morning's post. All in all, I think you've made a pretty good stab at it. And as for those drawings, they're absolutely cracking!

However, I hope you don't mind me taking the liberty of making a few judicious tweaks here and there. Actually, I've put the whole thing through my new 'EDIT-O-MATIC' machine, which should iron out a few bugs in the text. And design. And layout... To tell the truth, it's pretty much a new book altogether.

In fact, I'm sending you an exclusive preview of the plans for my invention - it's bound to bring this publishing game into the modern age! You could do worse than put your entire output through the works - just think of the improvements you could make at the flick of a switch. You'd hardly need to employ any real editors again!

Delighted to be of help, and can't wait to see the finished book!

Yours ever,

Wallace

Wallace

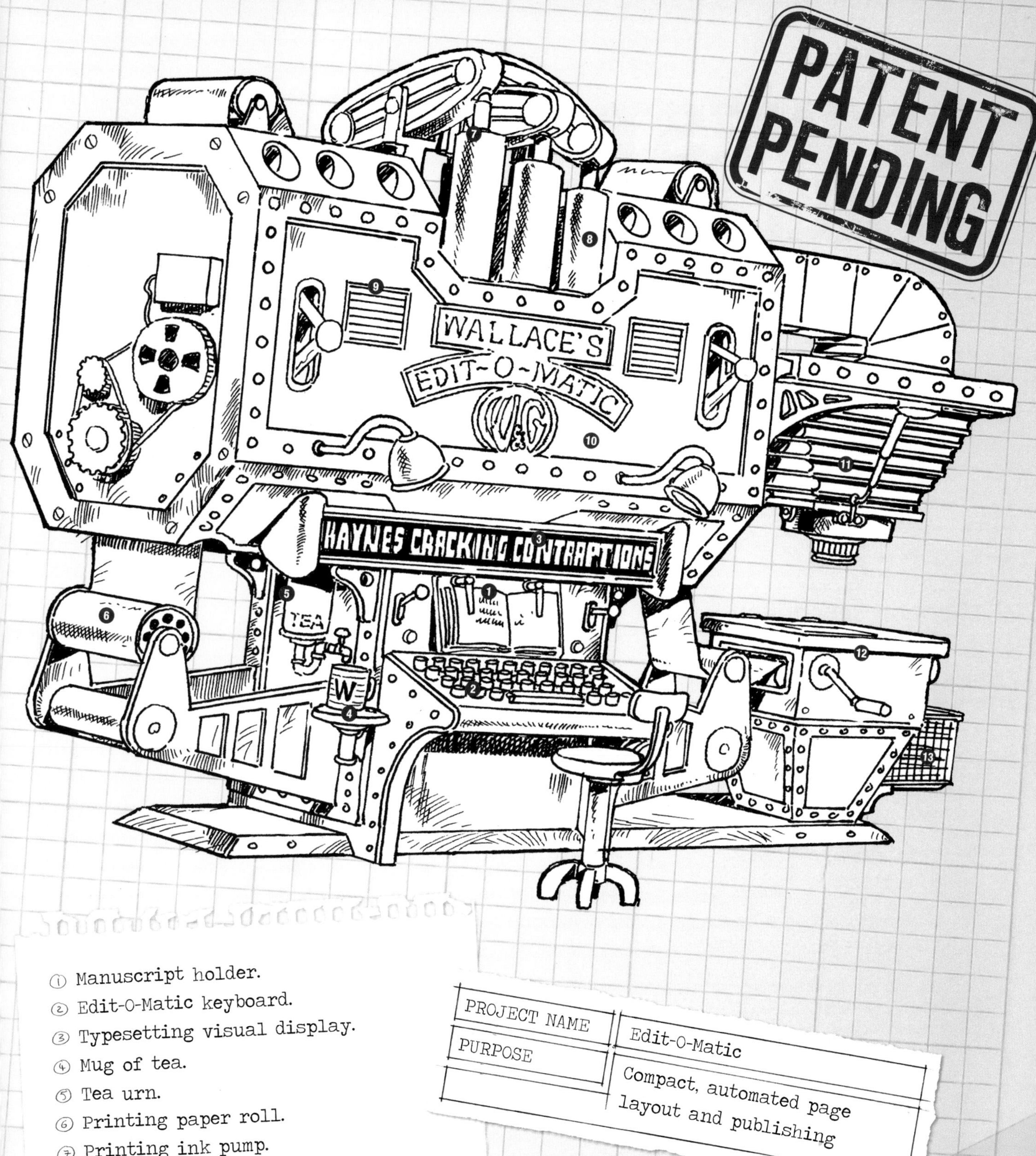

1. Manuscript holder.
2. Edit-O-Matic keyboard.
3. Typesetting visual display.
4. Mug of tea.
5. Tea urn.
6. Printing paper roll.
7. Printing ink pump.
8. Correction fluid pump.
9. Cooling vents.
10. Page layout mechanics housing.
11. Flat bed process camera adds photos and drawings to page layouts.
12. Pages are printed and trimmed, ready for binding.
13. Finished pages tray.

PROJECT NAME	Edit-O-Matic
PURPOSE	Compact, automated page layout and publishing

62 WEST WALLABY ST

Contents

General description

Wallace and Gromit live at 62 West Wallaby Street, in a spacious, double-fronted, semi-detached house. Downstairs there is an entrance hall, lounge, dining room, kitchen and pantry, and upstairs there are four bedrooms, one of which is used as a study, and a bathroom. Attached to the side of the house is a single garage, and there are attractive gardens front and rear.

Wallace is constantly inventing and many of his latest creations have been installed or put to use around the house. Over time these are often improved or replaced with new contraptions, and it's impossible to know what latest piece of ingenious machinery may be concealed behind a wall or above a ceiling.

Some of the more permanent (and more successful) contraptions include the Bed Launcher, which tips Wallace's bed and delivers him, via his trousers, to his chair at the breakfast table in the dining room below, where the Auto Dresser, Jam Ballista and Porridge Cannon are waiting ready to perform their duties each morning.

The 'Wash 'n' Go' launch system was devised when Wallace and Gromit started their window-cleaning business, and provided fast access to their motorcycle and sidecar. The system was later modified for use with their Austin A35 Van.

Other contraptions are called into action as necessary, and these include the Tellyscope, the Autochef, the Turbo Diner, the Snoozatron and the 525 Cracker Vac. All of them are prototypes and all of them need those few final tweaks to make them perfect.

Most of Wallace's inventing takes place in the unusually large cellar beneath the house. It is here that some of his largest and most spectacular contraptions have been designed and built, including the Rocket and the Knit-O-Matic.

From the outside, 62 West Wallaby Street appears to be just an ordinary suburban house and never really changes. That was until Wallace and Gromit started their Top Bun Bakery business, providing a complete 'Dough to Door' service to their customers. They installed a traditional windmill on the roof and converted the entire house into a large-scale bakery.

In the future, we might see even more dramatic changes to the house as Wallace's ideas and inventions get bigger and better.

WALI

PROJECT NAME	62 West Wallaby Street
PURPOSE	Home of Wallace & Gromit and their workshop (cellar)

62 West Wallaby Street cutaway

1. Garden path, leading to front gate and West Wallaby Street
2. Two pints of milk delivered daily
3. Lounge
4. Gromit's bedroom
5. Gromit's bed
6. Bookcase
7. Study
8. Wallace's bedroom
9. Wallace's bedside alarm system control box
10. Spare room (later redecorated and occupied by Gromit when Feathers McGraw takes his room)
11. Spare bed
12. Landing and stairwell
13. Part of the roof of the neighbouring house
14. Front door
15. Living room door
16. Cellar and workshop where many of Wallace's inventions are created
17. Launch system activation lever
18. Wallace's 'Wash-n-Go' job standby chair (shown in launch position)
19. Hydraulic lift raises Wallace's chair through the study and into launch position
20. Variable position launch chutes
21. Launch chute control box and gearing
22. Ladder attachment system
23. Spare ladder
24. Motorcycle engine auto-start boot
25. Overall suspension arms
26. Wallace's overalls
27. Telescopic sponge deployment arm
28. Garage cellar
29. Motorcycle and sidecar
30. Turntable
31. Hydraulic lift
32. Revolving garden pond (shown in 'GO' position)

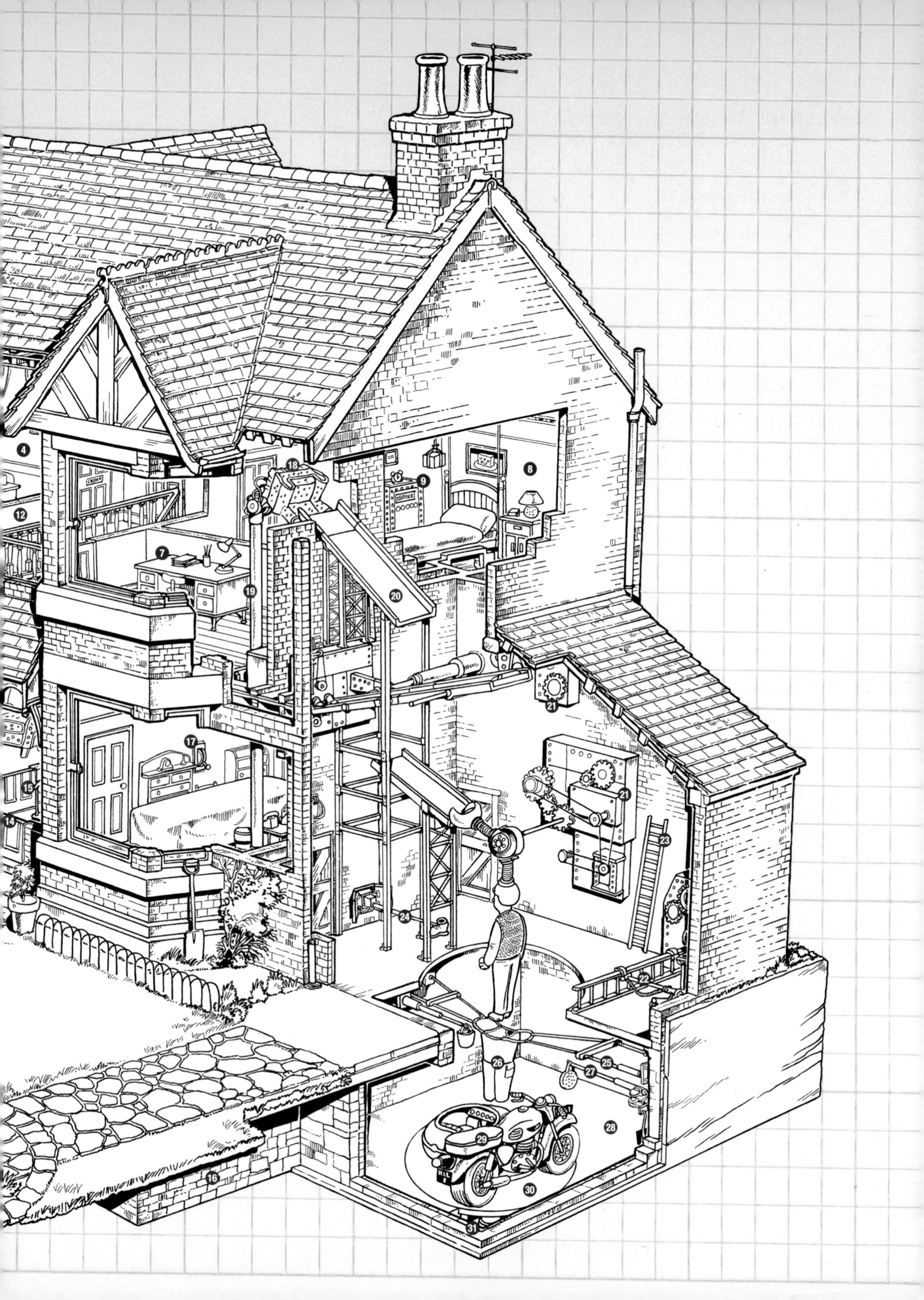
4
18
9
8
12
7
19
20
21
17
15
14
21
23
24
22
25
26
27
28
29
16
30
31

Wash 'n' Go launch system

One of Wallace's more ambitious inventions is the Wash 'n' Go launch system, which was installed when he and Gromit started their rapid-response window-cleaning business.

When a call is received from a distraught customer with dirty windows, they can be ready to go, along with all necessary window-cleaning equipment in less than a minute.

Wallace waits at the ready in his standby armchair. When the launch system is activated, the chair is lifted up through a hatch in the ceiling of the dining room, into the study above by a hydraulic lift. The chair is then tipped backwards and Wallace slides through the wall and down a series of chutes before being dressed in his overalls and delivered to his motorcycle and sidecar, which is automatically kick-started.

A ladder, sponge and bucket are then deployed and the Wash 'n' Go team are ready for action.

Launch sequence

❶ Once a call is received, Gromit pulls a lever, activating a hydraullic lift, which sends Wallace's standby chair (and Wallace) up into the study above.

❷ The chair tips backwards and a painting on the wall flips down, allowing Wallace to descend through the hole in the wall.

❸ Wallace slides head-first down the chute until his head comes to rest on a cushion at the other end.

❹ This part of the chute tips down and Wallace continues his descent feet-first.

❺ When he reaches the end, the chute tips again, and Wallace slides head-first again towards an awaiting motorcycle helmet.

❻ With his head gripped firmly in the helmet, Wallace is rotated clockwise to a vertical position.

❼ Wallace is turned 90 degrees to face the back wall of the garage, still with his head gripped in the helmet.

❽ Descending to his motorcycle, Wallace is dropped into his overalls while telescopic arms pass him his bucket and sponge.

❾ Wallace lands on the motorcycle, which is raised to garage floor level. It then rotates through 180 degrees and Gromit hops into the sidecar, having walked through a door from the pantry.

❿ A ladder is attached to the sidecar by extending arms, and the auto-start boot kick-starts the motorcycle.

⓫ The garage door slides up and over, and the motorcycle speeds over the garden pond – which revolves to reveal paving on the other side.

SERVICE
1
2
3
4
5
6
7
8
9
10
11

BED LAUNCHER

Contents

General description

The Bed Launcher provides the fastest possible means of getting Wallace out of his bed in the morning and to the breakfast table in the dining room, directly below his bedroom.

Wallace is woken by his alarm clock – although other devices may be employed if this doesn't work. He then uses the bedside pushbutton call system to tell Gromit, downstairs, what he wants, for example 'Breakfast'. This alerts Gromit, who throws a lever to activate the Bed Launcher.

The head end of the mattress is winched up by an electric motor, concealed above the ceiling, using a system of ropes and pulleys. The process is assisted by a counterweight, which falls down into the corner of the bedroom, pulling a rope and helping to lift the bed.

The arc made by the head end of the mattress is corrected by small slide wheels fitted at the foot end of the bed; this ensures that the winching ropes remain vertical during operation and increases the angle, and hence speed, of descent. The bedclothes are left untucked at the foot of the bed the night before so that Wallace can simply slide out once the mattress reaches the critical angle.

Simultaneously, a two-door hatch in the bedroom floor at the foot of the bed is opened by two rope winches. These are again operated by electric motors, this time mounted in the cavity between the floor of the bedroom and the ceiling of the room below.

Wallace slides out of the bed, through the hatch in the floor and straight into his trousers, which are suspended by their braces from four hooks located around the bottom edge of the hatch. This is a particularly ingenious part of the design since not only does it put on Wallace's trousers but it also helps to break his fall when he lands on his chair at the breakfast table in the room below.

As soon as Wallace lands on his chair, the Auto Dresser (see page 16 for full description) puts on the top half of his clothing (shirt, tie and tank top), while the sleeves are added by a separate device extending down from the ceiling.

The entire process takes a matter of seconds, which means that Wallace loses no time at the start of a busy day.

Further contraptions are then used to speed up or automate the serving of breakfast although none seem to work as effectively as the Bed Launcher.

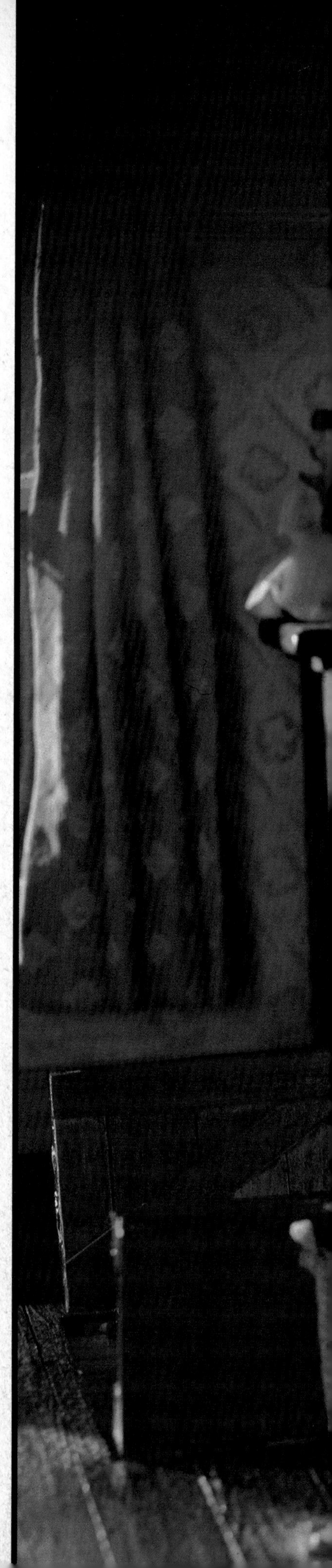

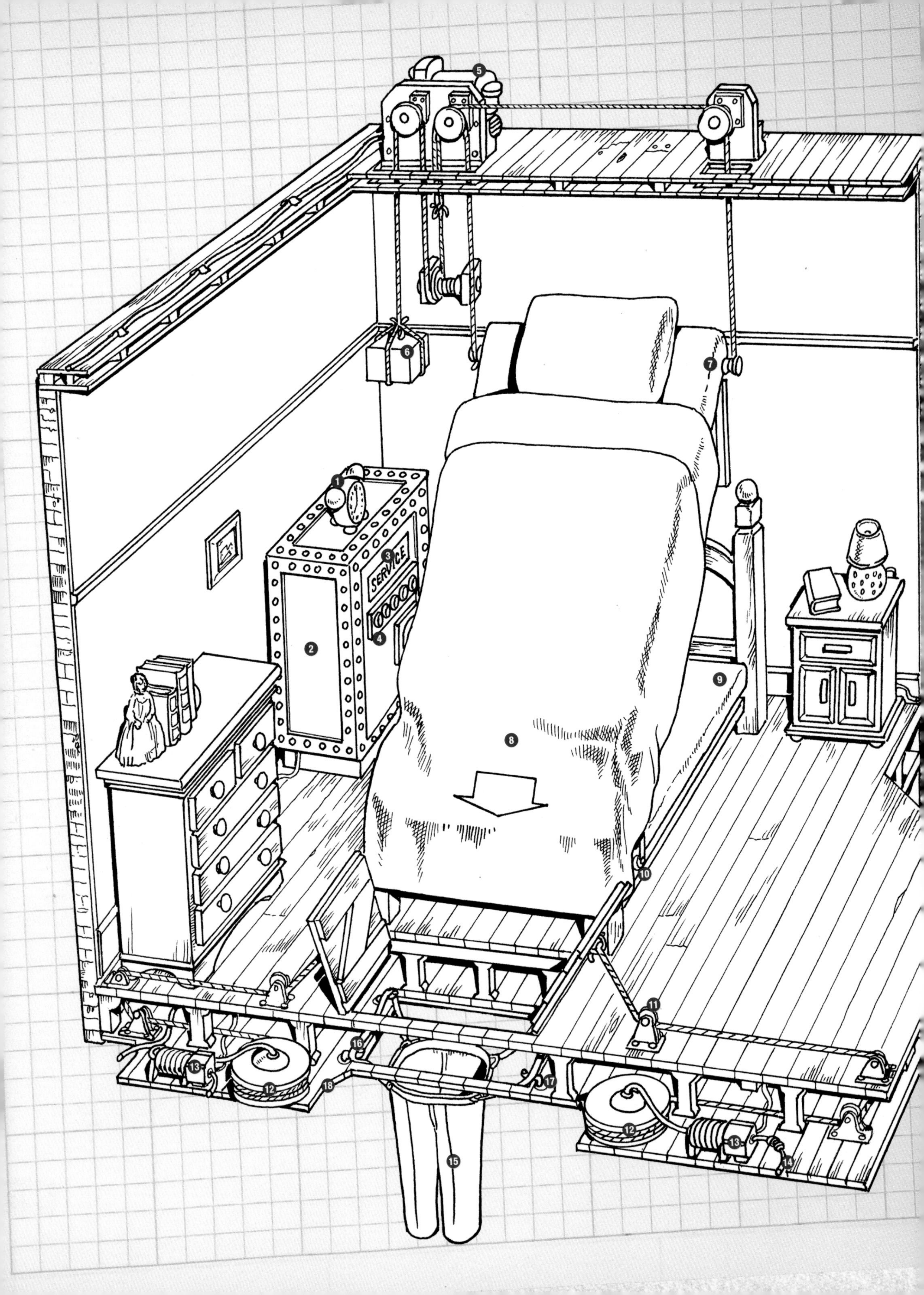
SERVICE

PROJECT NAME	Bed Launcher
PURPOSE	High-speed route from bed to breakfast via trousers

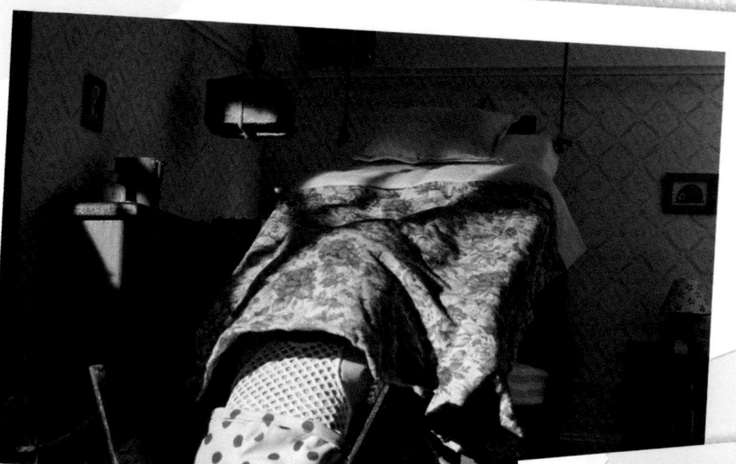

Bed Launcher cutaway

1. Alarm clock
2. Bedside call system control box
3. Visual display
4. Control box function buttons
5. Mattress lift electric motor
6. Mattress lift counterweight
7. Strengthened bed mattress
8. Blanket and eiderdown left untucked
9. Sprung base of bed
10. Mattress lift slide wheels allow foot end of mattress to slide back
11. Floor hatch pulleys and cord
12. Hatch cord winch reels
13. Hatch cord winch electric motor
14. Power cable
15. Wallace's trousers
16. Trouser braces placed on hooks the night before by Gromit
17. One of four trouser brace hooks
18. Dining room ceiling

Wallace slides out of bed and into his trousers . . .

. . . which also help to break his fall.

PROJECT NAME	Auto Dresser
PURPOSE	Automatic tank top, shirt and tie dressing machine

Auto Dresser – General description

The Auto Dresser's primary function is to dress Wallace in his shirt and tank top each morning. This is normally performed at the breakfast table after Wallace has been dropped through the ceiling and into his trousers from the bedroom above by the Bed Launcher. Shirt sleeves are pulled on to Wallace's arms by a separate device, and then the Auto Dresser adds the rest of the shirt, tie and knitted tank top to complete his outfit for the day.

The tank top and shirt combination is held and stretched open by two tank top holding bows. These are mounted on hydraulically actuated arms that are articulated and have complete freedom of movement, which is essential for achieving a comfortable fit. The arms are mounted on the control dome, which is able to rotate on top of the upper section of the Auto Dresser's body. The upper section can be raised and lowered in order to reach up and pull the clothes down over Wallace's head; the upper section also houses the motor that controls the arms as well as the rotation of the control dome.

The Auto Dresser is controlled by a computer, which contains several different dressing programs, and the user can change program using a dial on the control box at the top of the unit. The computer, control box, hydraulic arm actuators and gearing are all contained within the control dome.

The lower section forms the base of the Auto Dresser and contains the gearing to control the height of the upper section. The entire unit is mounted on caster wheels and driven by a second motor, also in the lower section, allowing it to move around the house as required by the current program. The Auto Dresser is also used in the cellar of 62 West Wallaby Street to assist in taking the latest item of knitwear produced by Wallace's Knit-O-Matic and putting it on the lucky recipient.

A hatch at the front of the lower casing allows access to the battery and for general maintenance of all internal components.

Auto Dresser cutaway

1 Program dial
2 Control box
3 Arm control circuit board
4 Good dress sensor dial
5 Tank top holder bows place clothes over Wallace's head; sleeves are added by separate sleeve attachment device
6 Tank top holder bow pivot
7 Arm hydraulics incorporating wiring for adjustable tank top holder bows
8 Computer controlling speed, direction and pre-programmed dressing functions.
9 Arm movement gearing
10 Rotating control dome
11 Rotation control gearing
12 Control dome rotation ball bearings
13 Top motor controls arms and rotation of control dome
14 Upper section maintenance key
15 Spare adjustment key
16 Upper section (in raised position)
17 Upper section height adjustment gearing
18 Height adjustment guide slot
19 Guide rollers
20 Lower section
21 Lower section maintenance hatch
22 Lower motor drives wheels, steering and height adjustment
23 Battery
24 Steering caster wheels driven by lower motor

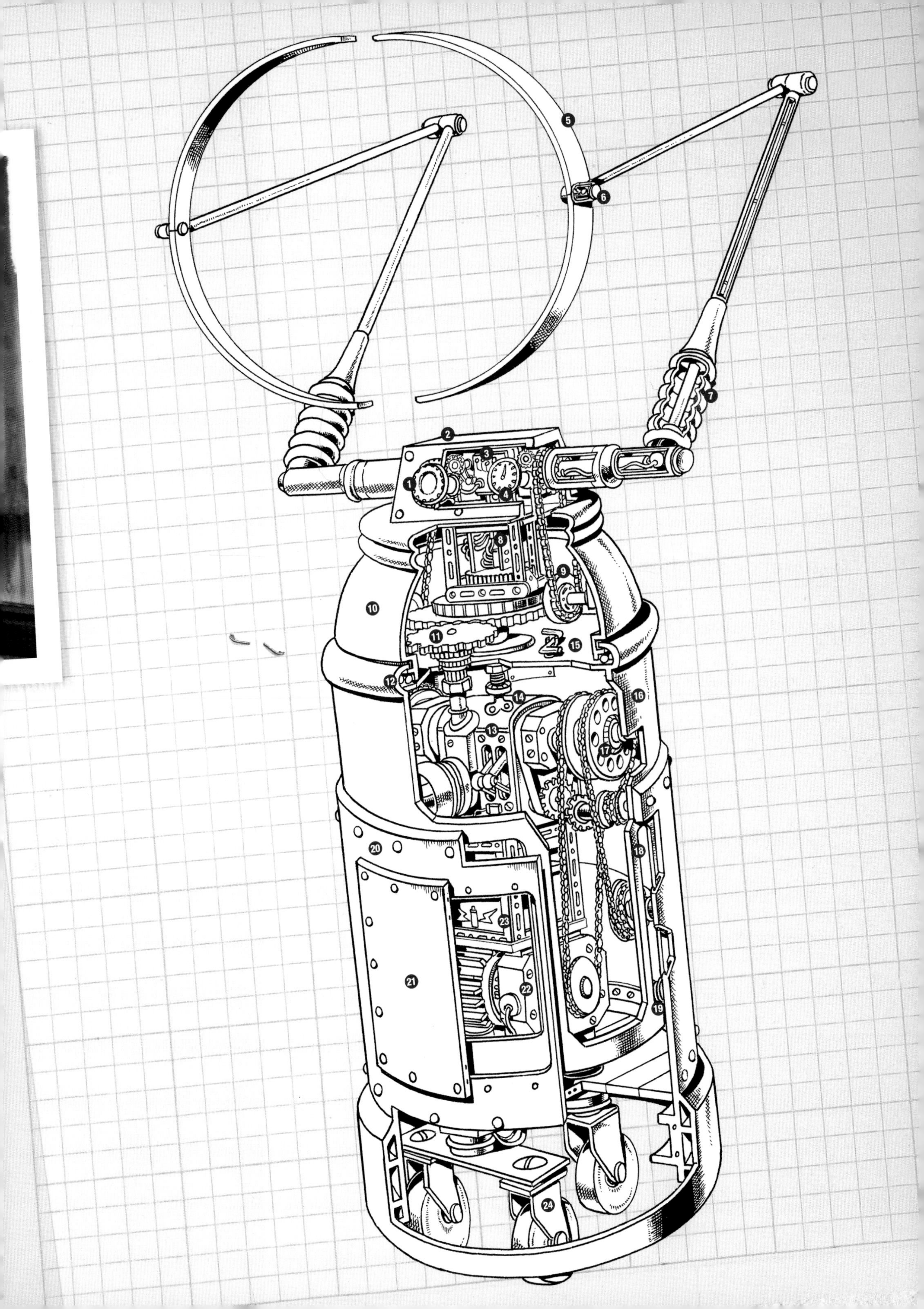

1
2
3
4
5
6
7
8
9
10
11
12
13
14
15
16
17
18
19
20
21
22
23
24

ROCKET

Contents

General description

Wallace wants to go on holiday, and where is there plenty of cheese? The moon of course! Since the easiest way to get to the moon is by rocket, Wallace and Gromit set about building one in the basement. The result is very much a home-made contraption, but the design ensures that passenger comfort is high on the agenda and that nothing is left to chance.

The rocket is of riveted construction and braced by six stanchions spaced equally around the inside of the egg-shaped hull. The internal space is split roughly in half with the 'top' half being the passenger cabin/living room. The lower half is almost entirely filled with the rocket engine, fuel, water and air tanks, fuse reel, an access ladder and various other equipment.

Access to the pressurised cabin is via a hatch from the outside. An additional hatch in the floor of the cabin provides access to the engine room and, finally, the engine room may be accessed directly from the outside via a maintenance access panel.

The propulsion system is a bespoke jet engine using a unique method of ignition. This consists of a length of fuse fed down through the rocket nozzle and

lit from the outside. The fuse is continuous and stored on a reel next to the engine so that a length may be deployed for the next lift-off.

A single operator console, situated in the passenger cabin, contains all the switches, levers and buttons necessary for navigation, steering and control of the rocket's engine. Further equipment includes an external view-scope, engine monitor headphones, an electric kettle and a fitted toaster.

Other than the control systems, the passenger cabin contains two armchairs, which serve to cushion Wallace and Gromit from the forces of acceleration during lift-off and to provide a comfortable place to relax during the journey, along with a range of various other items of furniture and home comforts.

Rocket cutaway

1. Hull support stanchion
2. Cabin lighting
3. Shelving for books and food supplies (mainly crackers for cheese)
4. Gromit's acceleration/ relaxation chair
5. Headphones for monitoring engine
6. Control systems console seat
7. Toaster
8. Clock
9. Observation porthole
10. Rocket control systems console (operated by Gromit)
11. External view-scope to monitor outside environment and to check altitude when landing
12. Pot plant to reduce carbon dioxide in cabin
13. Console internal mechanics
14. Console coolant tank
15. Rocket thrust standby control lever (handbrake)
16. Fuel tank
17. Fuel feed pipe
18. Fuel distribution/regulation valve
19. Rocket engine
20. Ignition fuse: stored next to the rocket engine and fed through the nozzle
21. Ignition fuse pulley system deploys an additional length of fuse for the return journey. However, it requires lighting from outside
22. Mouse trap
23. Engine room maintenance access panel
24. Ignition fuse cutter
25. Ignition fuse storage reel: holds up to 50 yards of fuse for further lift-offs
26. Cabin air tanks
27. Water tank (mainly for making tea)
28. Engine room light switch
29. Folding step ladder allowing access to Rocket from ground level
30. Ladder stowage compartment
31. Stabilising fin (one of three)
32. Access hatch to step ladder stowage compartment
33. Pressurised cabin access hatch
34. Knitting
35. Deckchair and spade
36. Wallace's acceleration/ relaxation chair
37. Observation porthole
38. Kettle
39. Camera
40. Holiday brochure
41. Radio
42. Bowl of fruit

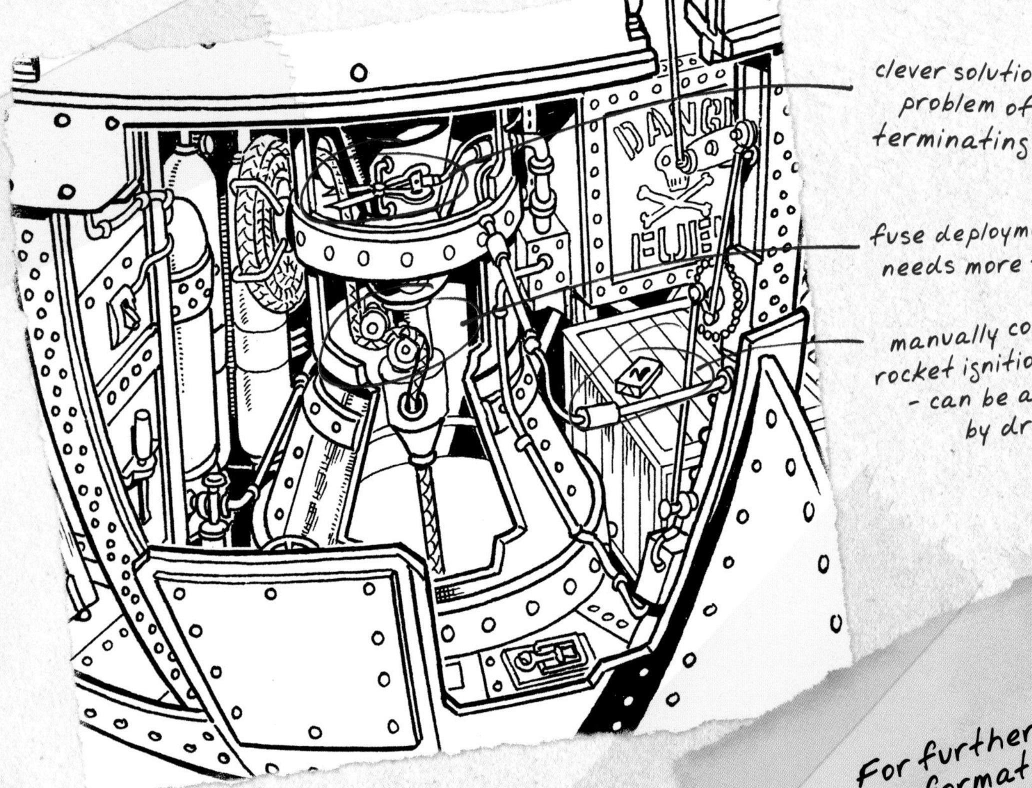

For further information on the Rocket's field test please see page 23

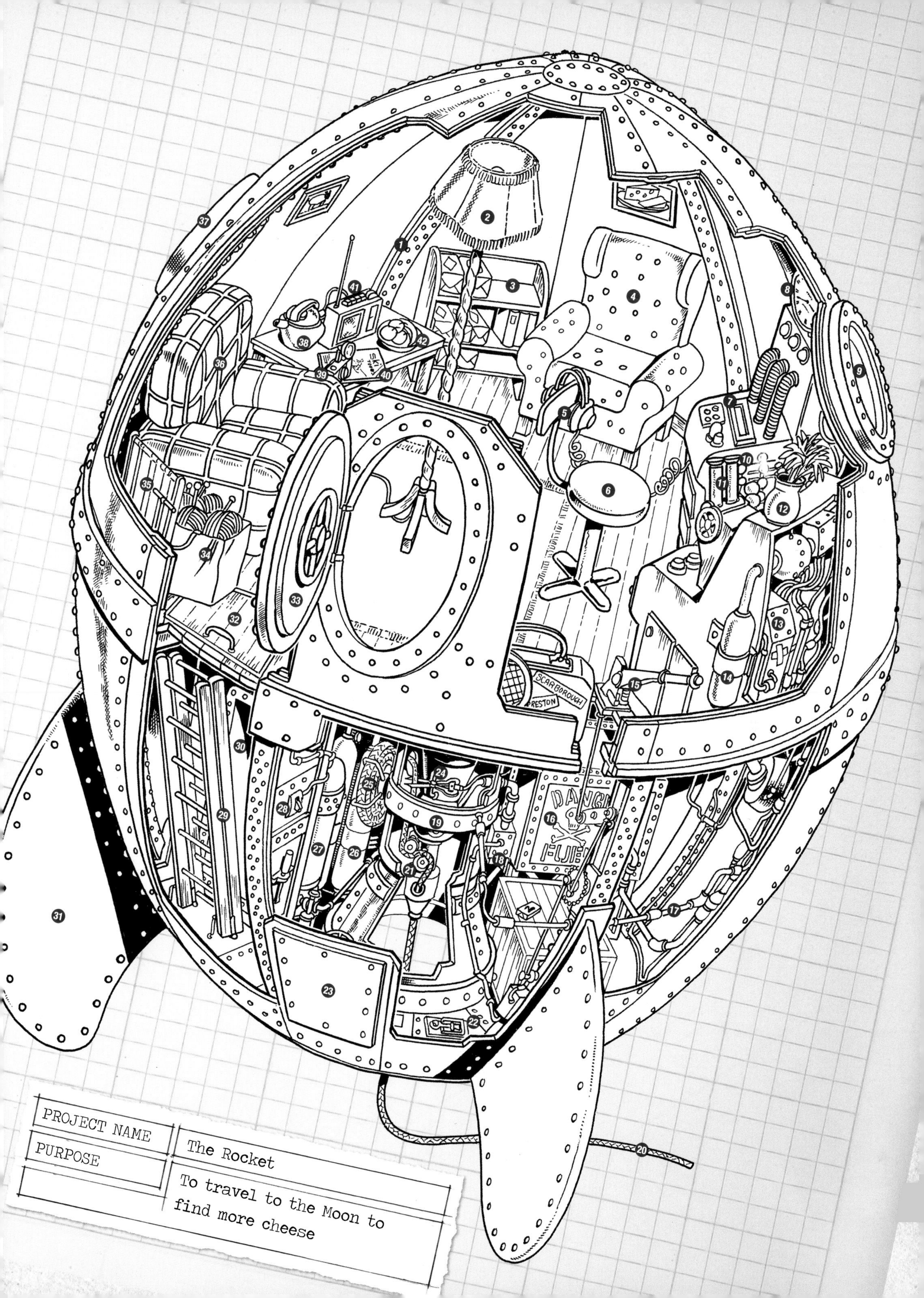

SKI
SCARBOROUGH
PRESTON
DANGER
FUEL
PROJECT NAME
The Rocket
PURPOSE
To travel to the Moon to find more cheese

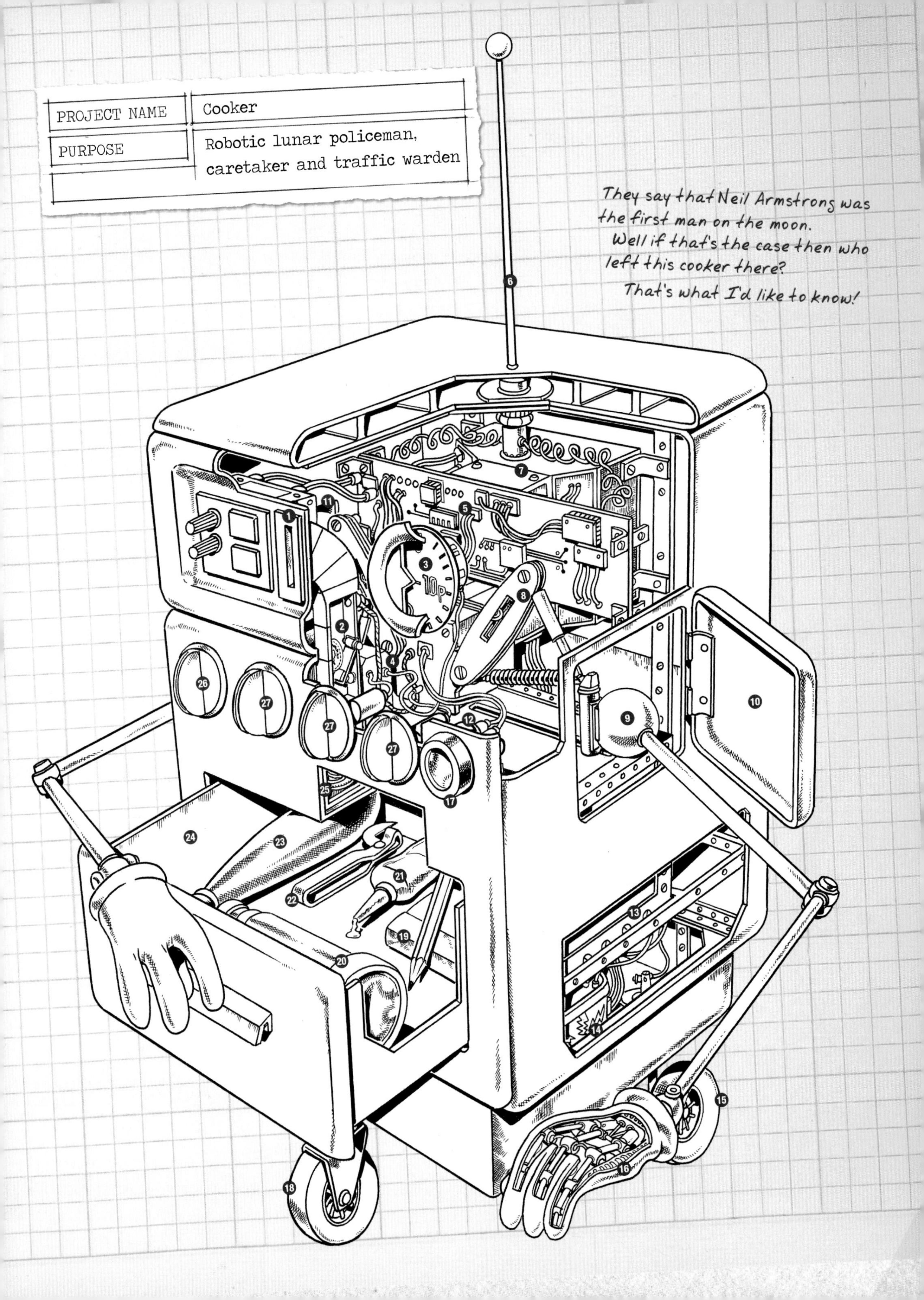
PROJECT NAME
Cooker
PURPOSE
Robotic lunar policeman, caretaker and traffic warden
They say that Neil Armstrong was the first man on the moon.
Well if that's the case then who left this cooker there?
That's what I'd like to know!
10p
1
2
3
4
5
6
7
8
9
10
11
12
13
14
15
16
17
18
19
20
21
22
23
24
25
26
27

During their trip to the moon to sample the 'lunar cheese', Wallace and Gromit discover a machine, which appears to be a domestic gas cooker mounted on trolley wheels with an antenna. The front cover houses a coin slot, into which Wallace inserts a coin, but nothing happens at first. However, as soon as Wallace and Gromit walk away, the coin mechanism operates and the machine comes alive.

Immediately, flaps in the side panels open to reveal robotic arms complete with articulated joints and hands. The Cooker begins to clear up the remains of Wallace and Gromit's picnic, but on finding one of their discarded holiday brochures about skiing its thought visualisation processor stirs into action and it begins to 'daydream'.

A drawer in the Cooker's lower casing provides storage for various implements including a dustpan and brush, a notepad and pencil, a truncheon, a tin opener and a telescope, which it uses to search the landscape and discover Wallace and Gromit's Rocket (illegally parked and leaking oil). The Cooker gives the Rocket a parking ticket before searching again for the culprits, who it discovers tucking into more moon cheese. Racing across the moon surface, the Cooker grabs the truncheon from its drawer and is about to strike Wallace over the head when, fortunately, the coin-op mechanism times out and the machine comes to a sudden halt. Wallace takes the truncheon and inserts another coin into the Cooker. Once more the machine springs to life but realises that Wallace and Gromit are on their way back to the Rocket and Earth...

The Cooker reaches the Rocket before it can take off and gains entry through the lower hull using its tin opener. Unable to see in the dark, the Cooker strikes a match, causing an explosion and ignition of the Rocket's engine. The Cooker is thrown from the Rocket, but manages to tear off two strips from the steel hull. Although unable to travel to Earth, the Cooker fashions the steel strips into makeshift skis, which it uses to ski up and down the lunar landscape.

Cooker cutaway

1. **Coin slot. Inserting a coin activates timer-control mechanism**
2. **Memory activation solenoid, triggered by coin insertion**
3. **Timer-controlled price display, linked to and triggered by Memory activation timer**
4. **Memory activation timer**
5. **Primary functions circuit board: controls speed and direction of travel, arm movement and visual processing**
6. **Thought visualisation processor antenna**
7. **Memory bank and thought processing circuits**
8. **Arm control hydraulics: when not in use, the arms are retracted and stored behind the coin slot**
9. **Left arm shoulder swivel joint extended to side hatch doorway to allow maximum movement**
10. **Left arm hatch**
11. **Right arm hatch**
12. **Visual processor linked to memory bank via primary functions circuit board**
13. **Drawer housing (originally the main oven)**
14. **Battery**
15. **Rear left wheel**
16. **Left hand manipulation and control actuators**
17. **Visual processor camera lens**
18. **Front steering wheel**
19. **Notebook and pencil**
20. **Telescope for long-distance visual enhancement**
21. **Lunar surface instant repair kit (quick-drying glue)**
22. **Tin opener**
23. **Anti lunar intruder device (truncheon)**
24. **Main oven door replaced by drawer unit**
25. **Used coin storage container**
26. **Thought visualisation processor 'tuning' control knob**
27. **Original oven control knobs (redundant)**

JAM BALLISTA

Contents

Watching the toast pop up gave me a cracking idea

General description

The Jam Ballista is a relatively straightforward contraption, designed for one reason: to get jam on to a piece of toast the moment it pops out of the toaster. A toaster pop-up synchronisation terminal ensures that the operation is timed to perfection.

The device itself uses a dipping spoon, which is loaded with the user's jam of choice directly from the pot. This simple design means that there are no internal workings to become gummed up with jam.

The catapult spring is extended and held in tension by a jointed spar, located in the centre of the spring coils. The spar is extended by an activation motor operating via a rocker assembly in the base of the Jam Ballista. When the activation (or firing) button is pressed, the current to the motor is cut off, releasing the spar from its primed state and allowing the catapult spring to contract. The spoon pivots in the support structure and comes to a sudden stop as the lower end of the spar lands on its shock-absorbing pad, causing the jam to be launched.

A dial on the control panel allows the user to pre-set the amount of tension loaded into the catapult spring, thereby allowing a degree of control over speed and trajectory.

THICK PORRIDGE MIX
News
JA

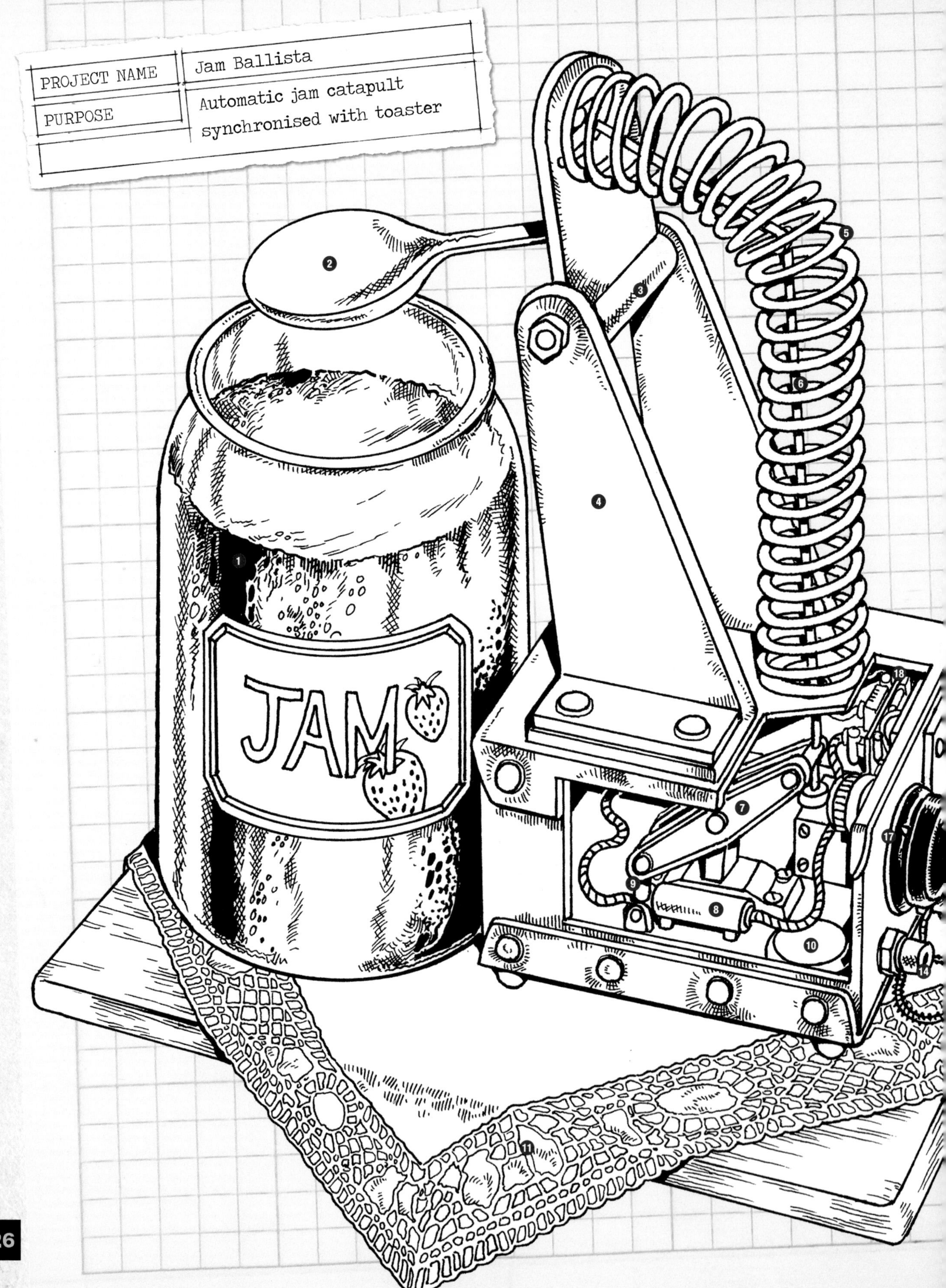
PROJECT NAME
Jam Ballista
PURPOSE
Automatic jam catapult synchronised with toaster
JAM
1
2
3
4
5
6
7
8
9
10
11
14
17
18

Jam Ballista cutaway

1. Open pot of jam (strawberry shown but other flavours optional)
2. Jam dipping spoon
3. Catapult pivot
4. Hardened steel support structure
5. Catapult spring
6. Spring extension spar
7. Activation motor rocker
8. Spring extension spar activation motor
9. Spring tension sensor
10. Spar shock-absorber landing pad
11. Doily to reduce vibration
12. Activation (firing) button
13. Cable to activation button
14. Toaster synchronisation terminal
15. Toaster synchronisation cable
16. Power switches
17. Spring tension pre-set dial
18. Control circuitry

Ex-NASA

TECHNO TROUSERS

Contents

General description

The Techno Trousers were designed by NASA to provide automated personal transport for astronauts during surface landings and other extra-vehicular activities, such as the repair of spacecraft. Not only are they fully automated, and controllable by remote, but they are also pre-programmable, which makes Wallace think that they will make the ideal dog-walking companion (and birthday present) for Gromit. Perhaps not surprisingly, Gromit isn't quite so impressed.

When used for transport, the operator (or 'pilot') sits on a padded saddle in the upper 'waist' section of the Techno Trousers. Each foot is located inside a leg of the device and secured to a foot platform by a toe clip. With the rise and fall of the legs, the foot platforms are also raised and lowered hydraulically to compensate, maintaining stability and maximising comfort for the pilot who experiences only a simple back-and-forth movement of the legs.

Among the Techno Trousers' many features are their ability to walk up vertical walls using switchable vacuum and magnetic field generators located in the soles of the integrated all-terrain boots. This system allows the boots to cling to both metallic and non-metallic surfaces, permitting the Techno Trousers to scale virtually any obstacle, and even walk upside-down. This function proves extremely useful while Gromit redecorates his bedroom but also leads to trouble when the Techno Trousers fall into the wrong hands...

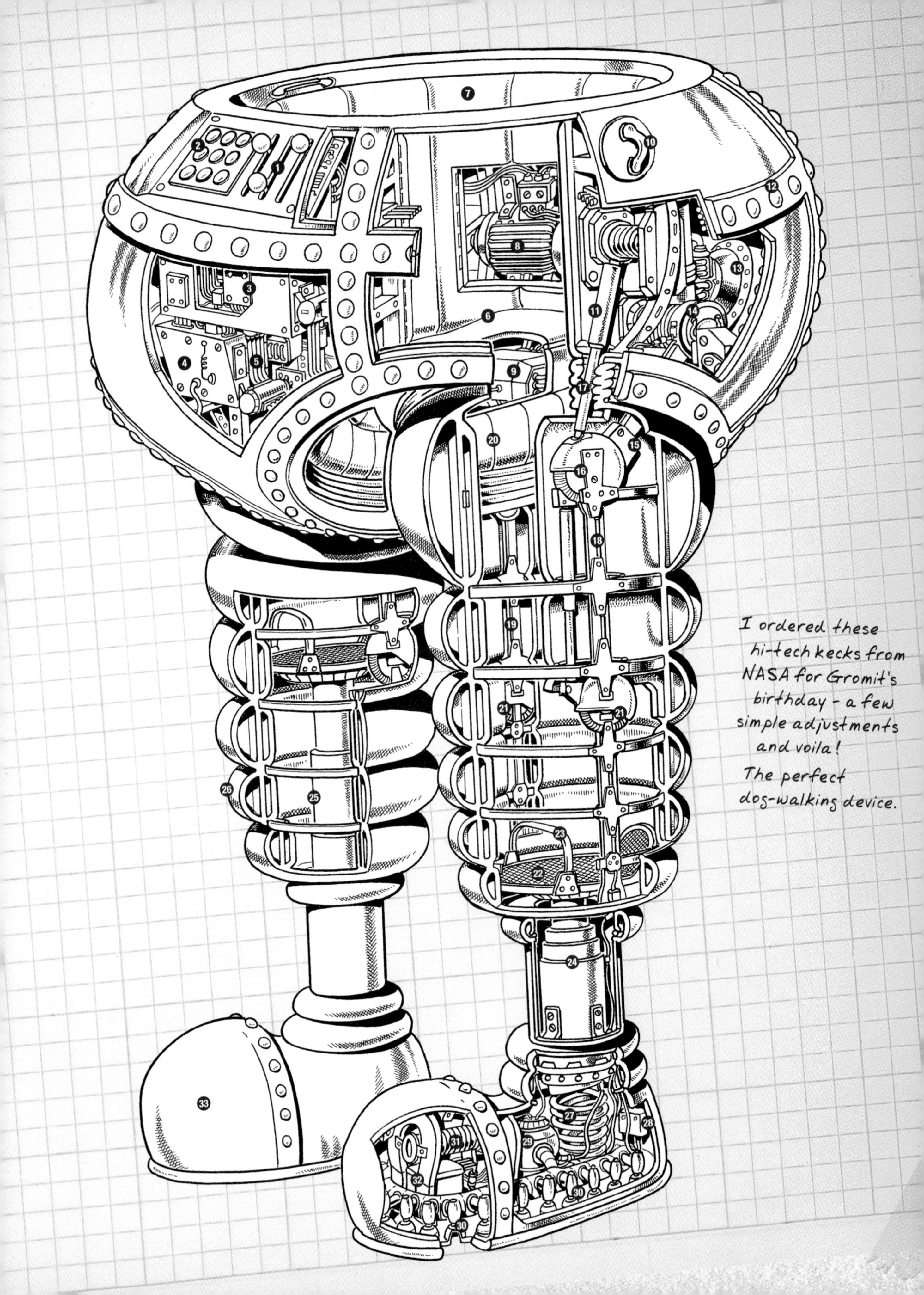
I ordered these
hi-tech kecks from
NASA for Gromit's
birthday - a few
simple adjustments
and voila!
The perfect
dog-walking device.

Techno Trousers cutaway

1. Operator's controls (designed to be manipulated with spacesuit gloves)
2. Automatic pre-programming controls
3. Control system computer
4. Remote-control receiver allows Techno Trousers to be controlled from a landing ship or base camp if operator is incapacitated
5. Control lines to motors
6. Operator's saddle and battery compartment cover
7. Internal padding to cushion impact and movement
8. Primary electric motor, powered by rechargeable battery or optional extension lead (top speed: 'Walk Factor 10')
9. Rechargeable battery
10. Techno Trousers port-side loading hook
11. Forward/reverse leg movement actuator
12. NASA-grade rivets
13. Rear external power supply and recharging socket
14. Left leg secondary motor
15. Leg speed gearing module
16. 180° hip motor
17. Flexible leg-joint seal
18. Outer internal leg sub-frame
19. Inner internal leg sub-frame
20. Flexible internal padding prevents operator's legs from chaffing
21. Knee joint motors
22. Operator's foot platform (fully compressed)
23. Toe clip
24. Foot clamp variable height adjustment hydraulics
25. Operator's foot platform (fully extended)
26. Flexible rubberised outer skin permits maximum leg movement and provides protection from inclement weather
27. Footfall impact absorber
28. Foot electronics crossover and junction box
29. Switchable vacuum generator allows Techno Trousers to scale vertical surfaces and walk upside-down.
30. Sole vacuum tubes
31. Gravity field repulsor coil
32. Localised field generator magnets allow boots to grip metallic surfaces
33. Armoured all-terrain boots

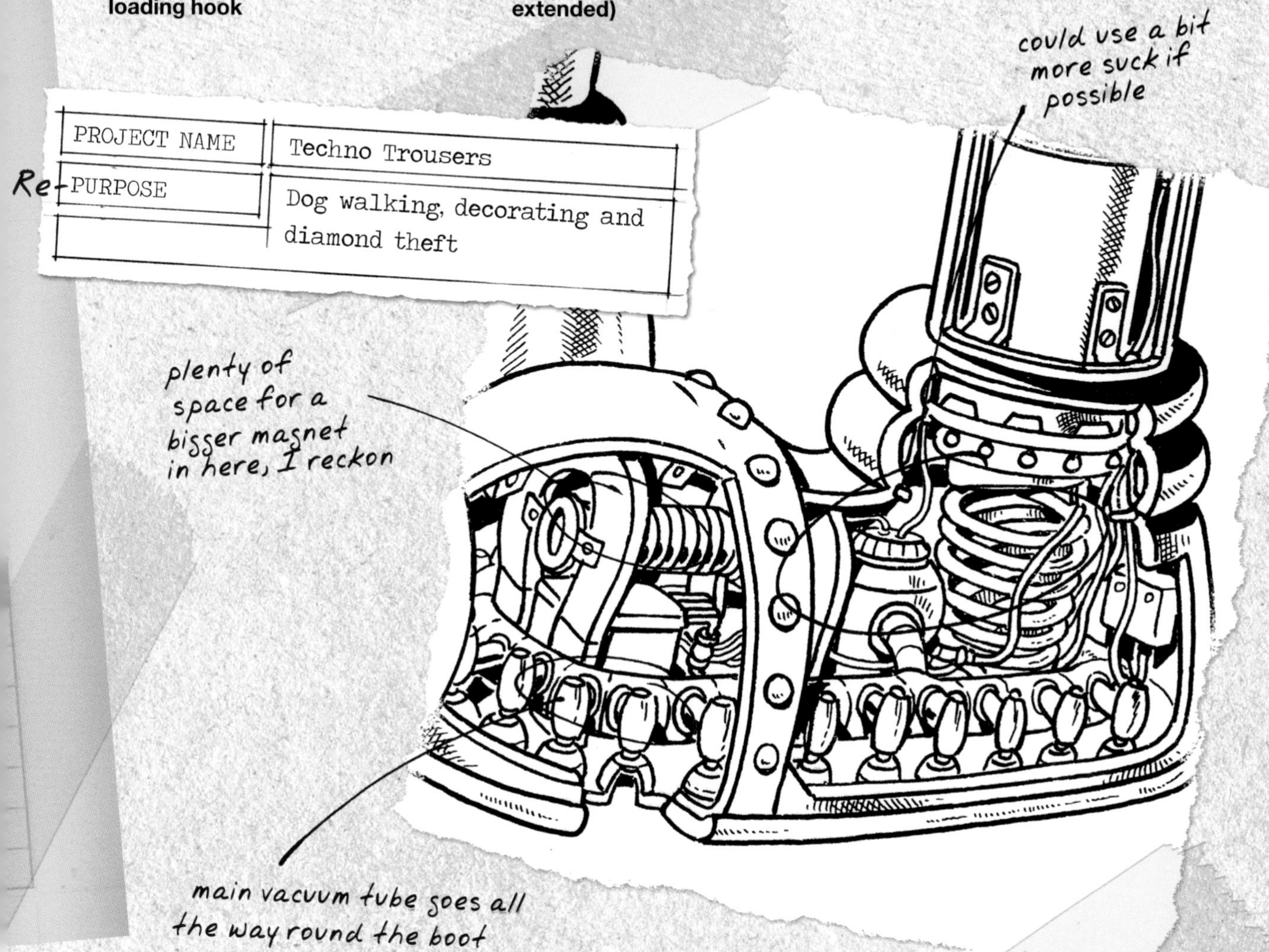

When notorious criminal Feathers McGraw rents a room at 62 West Wallaby Street, it doesn't take him long to realise the full potential of the Techno Trousers. He devises a cunning plan to steal a large and extremely precious diamond from the Town Museum using Wallace as his unwitting accomplice and the Techno Trousers with some special modifications of his own.

Firstly, the operator's controls on the front of the upper section are removed and a tamper-proof panel cover is put in their place. This means that the Techno Trousers can now be controlled only via a new remote-control unit, which has been specially configured by Feathers McGraw.

Secondly, Wallace (by now exhausted and fast asleep after a long day of extensive 'testing') is fitted with a helmet in which is stowed a Diamond Grabber claw. This is operated (also by remote) while Wallace is hanging upside-down directly above the diamond in the museum. The grabber claw is deployed through a pair of opening panels in the top of the helmet by a cable winch; it then grabs the diamond, retrieving it for Feathers McGraw who is waiting safely outside.

Meanwhile, Gromit discovers the true identity of the new lodger and confronts him upon his return to West Wallaby Street. Following a struggle, Feathers McGraw is finally captured and towed anonymously to the police station by the Techno Trousers.

The scoundrel had been tinkering and right under our noses too!

I had a very tiring day in the WRONG trousers

Feathers McGraw controls cutaway

1. Operator's control switches have been removed, and moved to the remote-control unit
2. Pre-programming controls have been removed and rewired. All functions have been duplicated on the remote control reusing the original control switches
3. Original computer memory store now rewired and reprogrammed
4. New remote-control radio link
5. Remote-control receiver has been overridden, with wiring re-routed to new remote-control radio link
6. Re-routed control lines to motors
7. Tamper-proof panel cover ensures that Techno Trousers can be controlled only via new remote control
8. Safety braces (originally installed by Gromit)
9. Remote-control unit
10. Remote controls duplicate the disconnected operator's controls on the Techno Trousers
11. Pre-programming control system transferred to, and reprogrammed by, the remote unit
12. Neck strap

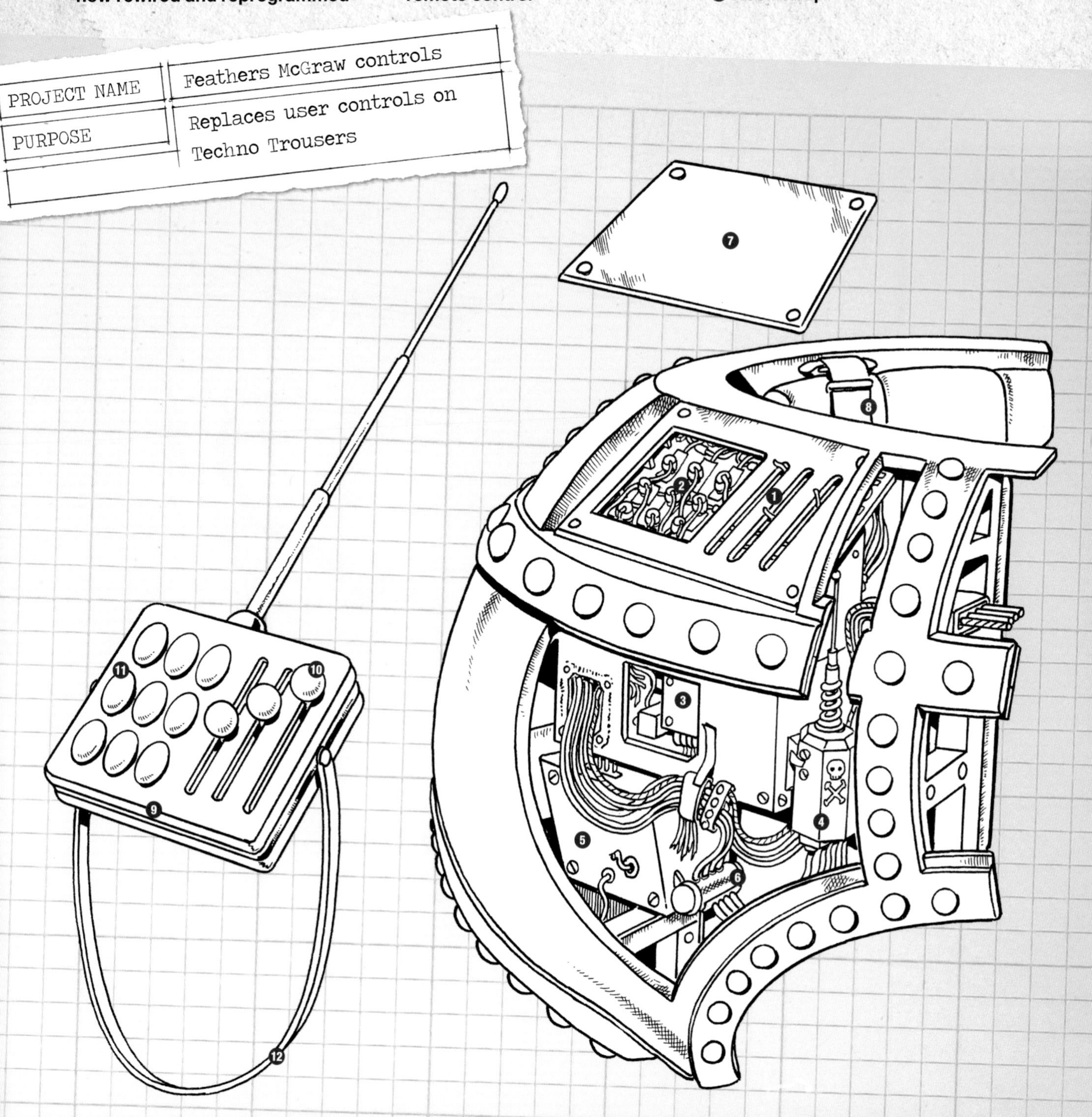

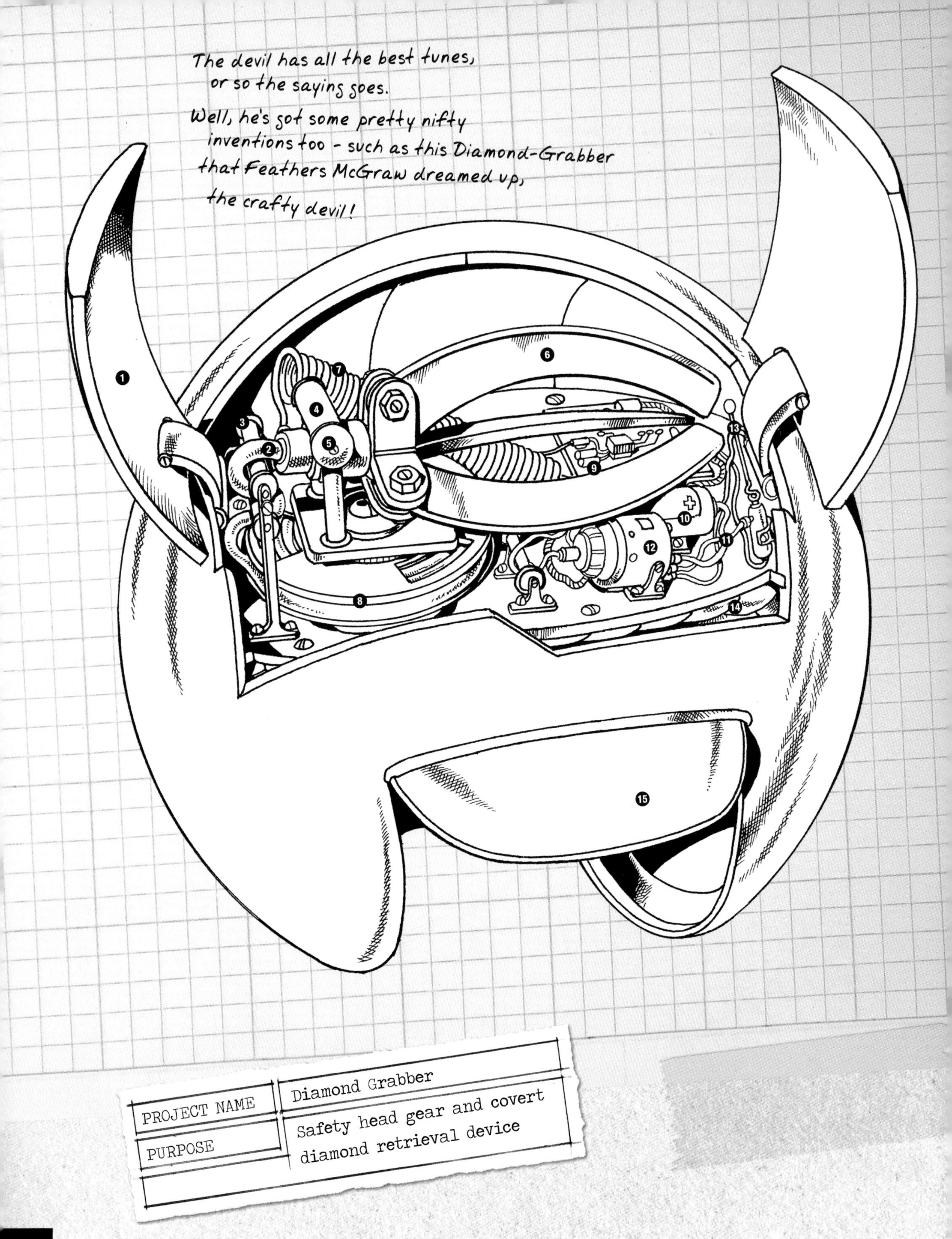
The devil has all the best tunes, or so the saying goes.
Well, he's got some pretty nifty inventions too - such as this Diamond-Grabber that Feathers McGraw dreamed up, the crafty devil!
1
2
3
4
5
6
7
8
9
10
11
12
13
14
15
PROJECT NAME
Diamond Grabber
PURPOSE
Safety head gear and covert diamond retrieval device

Diamond Grabber cutaway

1. Grabber deployment panel
2. High-strength grabber deployment cable
3. Cable deployment guide rollers
4. Clamp roller
5. Retaining clamp and roller locks grabber inside the helmet when not in use, and ensures smooth deployment when grabber is activated
6. Three-pronged diamond grabber shown in stowed position (best results are achieved when helmet is used upside-down)
7. Grabber electronics control cable
8. Cable spool rewinds grabber back into helmet after use
9. Electronics and control circuit board
10. Battery
11. Deployment panel hinge control wiring
12. Electric motor
13. Antenna links grabber to new Techno Trousers remote control
14. Internal helmet comfort padding
15. Visor

PORRIDGE CANNON

Contents

General description

It is probably best to think of this contraption as a high-powered air gun, which is designed to fire porridge either ready-mixed from a tank or made from dry porridge oats and water using the on-board mixer. The domestic version (shown here) is mounted on a substantial metal base unit, which is motorised and steerable to allow for convenient positioning at the breakfast table. A modified version, the 'Sud Cannon', is fitted to the sidecar of the motorbike and used for Wallace and Gromit's window-cleaning business (see page 56).

Ready-mixed or 'wet' porridge is stored in a tank, held high behind the cannon, from where it is drawn through a flexible pipe to the mixer and then into the porridge firing chamber. The chamber is then compressed by a high-pressure pneumatic ram, driven by air from a pump within the base unit. This compacts the porridge within the chamber and forces it from the barrel of the cannon at high speed. When used in 'automatic fire mode' this whole operation can be repeated several times a second and continue indefinitely (or until the porridge tank is empty).

A control panel on the front of the base unit features a master on/off switch together with various programming controls and status lights. These may be used to control the speed and aim of fire plus the consistency of the porridge itself. Finally, a remote-control start/stop button is connected by a cable to the porridge cannon for added convenience.

CRACKERS

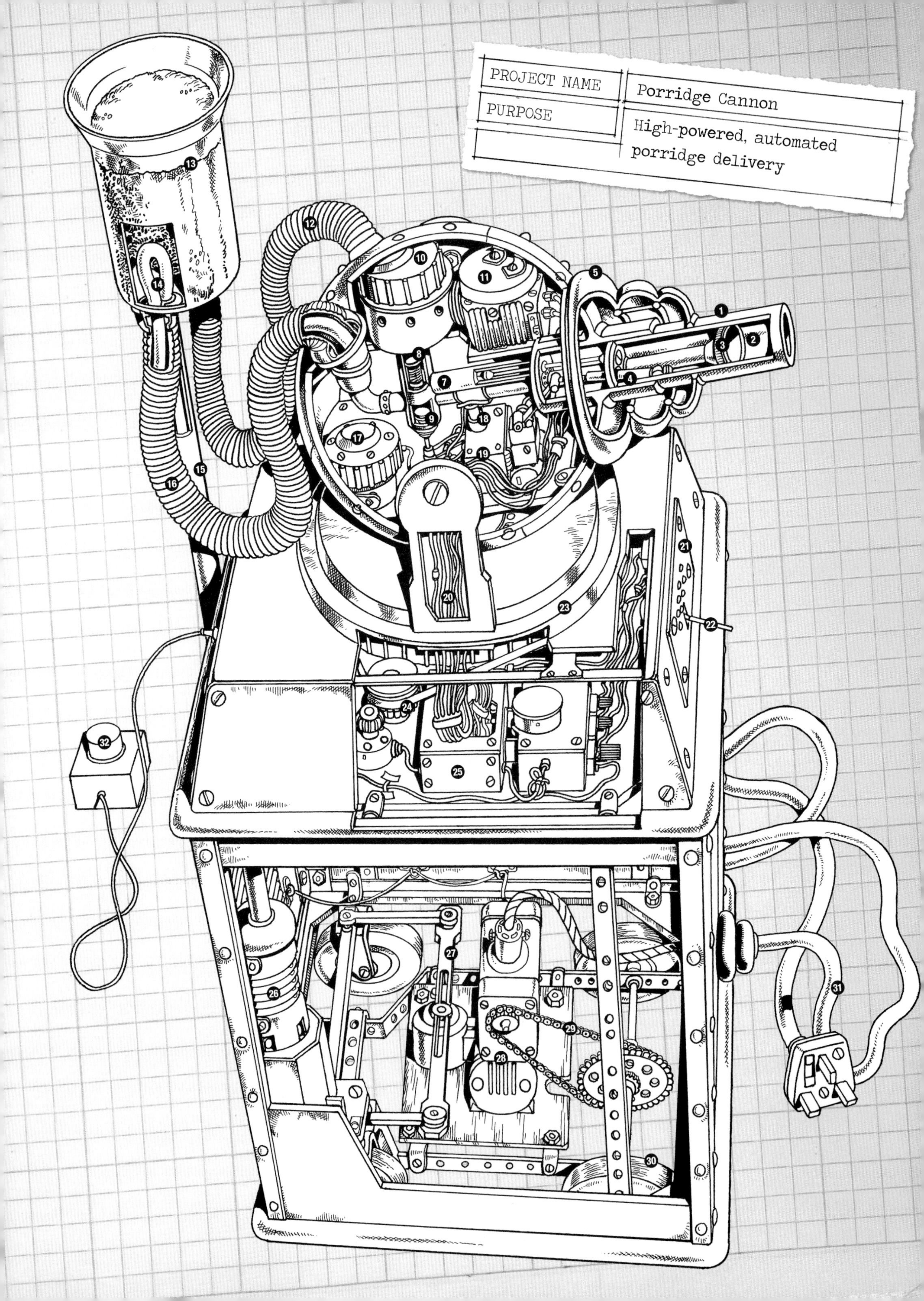
PROJECT NAME
Porridge Cannon
PURPOSE
High-powered, automated porridge delivery
1
2
3
4
5
7
8
9
10
11
12
13
14
15
16
17
18
19
20
21
22
23
24
25
26
27
28
29
30
31
32

Porridge Cannon cutaway

1. Automatic cannon barrel pump
2. Porridge firing chamber
3. Porridge nozzle
4. Pneumatic ram
5. Cannon barrel recoil dampener
6. Piston
7. Air cylinder
8. Primary valve
9. Secondary valve
10. Porridge mixer: ensures desired consistency with either ready-mixed or dried porridge
11. Mixer motor
12. Porridge pipe
13. Ready-mixed (wet) porridge tank
14. Flexible air pipe
15. Air pipe and porridge container support stand
16. Outer hose carries water to mixer
17. Water pump: used when cannon is fitted to motorbike sidecar (see page 56)
18. Electrically controlled trigger
19. Trigger control timer and motor
20. Control wiring
21. Programming controls
22. Master on/off switch
23. Cannon turntable
24. Turntable drive belt
25. Control unit (programmable)
26. Air pump
27. Steering control bar
28. Dual electric motors operate wheels and air pump
29. Front-wheel chain drive
30. Drive wheels
31. Power cord and plug *
32. Remote control start/stop button

* Shaun chewed the power cable so it's a bit iffy

Maybe just another couple of tweaks to the targeting too

KNIT-O-MATIC

Contents

General description

The Knit-O-Matic is one of Wallace's greatest inventions and takes up almost the entire cellar below 62 West Wallaby Street. It is a completely automated machine for washing, drying and shearing sheep, spinning and dying the wool, and then using it to knit a garment. Putting it simply, a sheep goes in at one end and a knitted jumper comes out at the other end.

The sheep starts in the Auto Wash tub where it is gently scrubbed by two hydraulically operated washing sponges to remove all dirt and odours from the wool. An extendable suction tube is then lowered and the sheep is sucked up by a powerful fan to the Auto Dry, where it is dried by drying fans before being passed along the clean sheep duct to the next stage in the process. Although part of the full Knit-O-Matic, the Auto Wash and Auto Dry machines may be controlled separately by a control console, which is responsible only for these functions.

Now that the sheep is clean and dry, it is gravity-fed into the Auto Shaver unit* via a vertical tube, where a soft landing is provided by an impact-reducing trampoline. Six articulated shearing arms each carry an electric razor to ensure fast and efficient shearing of the sheep. As soon as the sheep is shorn it is ejected from the Auto Shaver unit via a duct in the side wall and out through an exit hatch. In the meantime, suction fans gather the shorn wool and pass it to the wool-processing drum (carding drum) where it is carded (straightened) ready for spinning. As the spinning cords exit the Auto Shaver they are dyed in a range of colours as required by the machine's current program.

The wool cords are then spun into yarn using an automated spinning machine from where it is fed into the Auto-knit machine, which can be programmed to create a range of fashionable knitwear.

The operator has overall control of the Knit-O-Matic using the main system control console. While wool dying, spinning and knitting functions can be programmed, a single dial is used to choose either a wash only for the sheep or a light, medium or close shave. Unfortunately, this variable-cut shaving control dial has proved to be somewhat unreliable, sometimes switching automatically to 'close shave' when the Knit-O-Matic develops a system fault.

* see p45 for detail

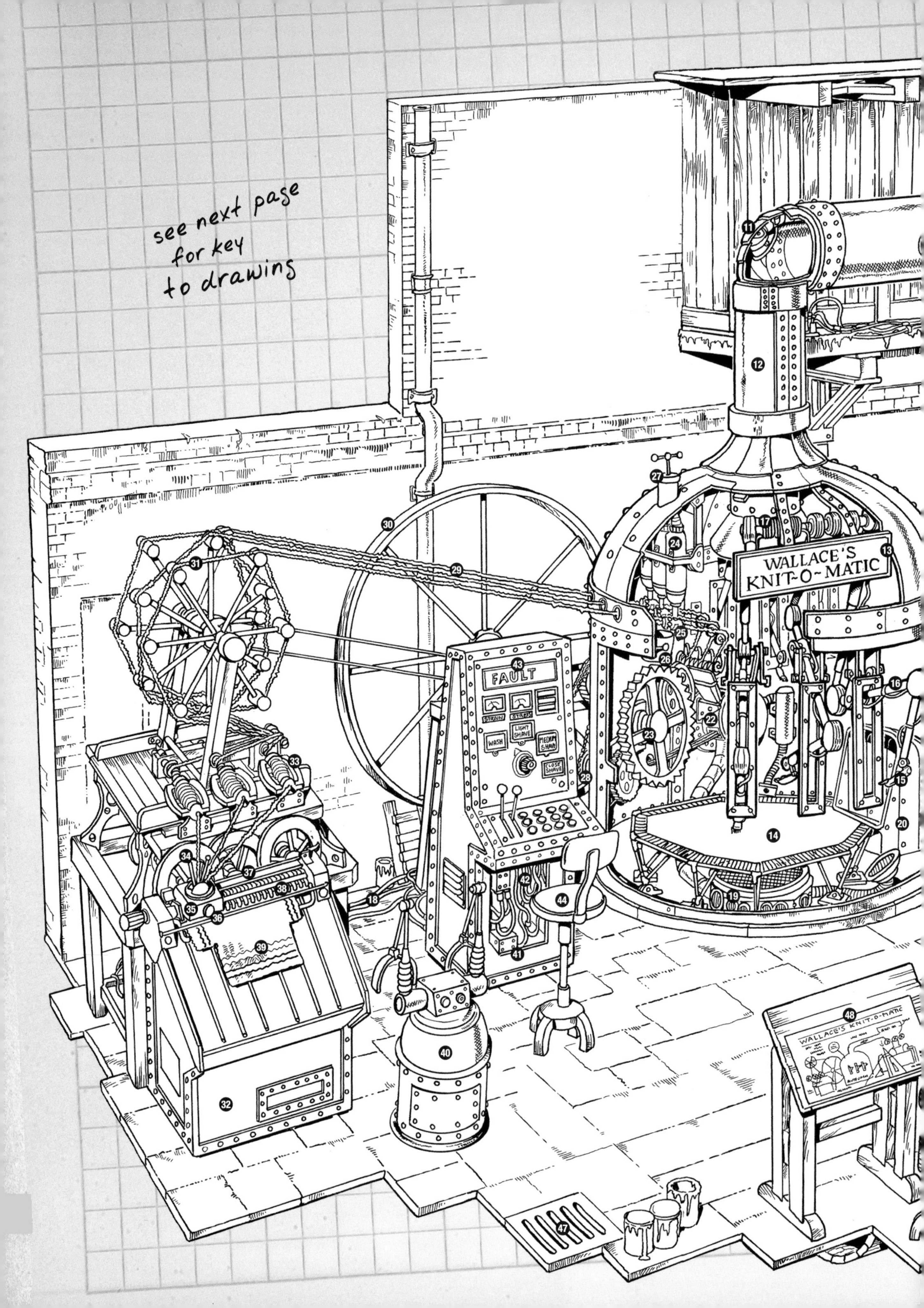
see next page for key to drawings
WALLACE'S KNIT-O-MATIC
FAULT
STRAIN
STRESS
WASH
LIGHT SHAVE
MEDIUM SHAVE
CLOSE SHAVE
WALLACE'S KNIT-O-MATIC

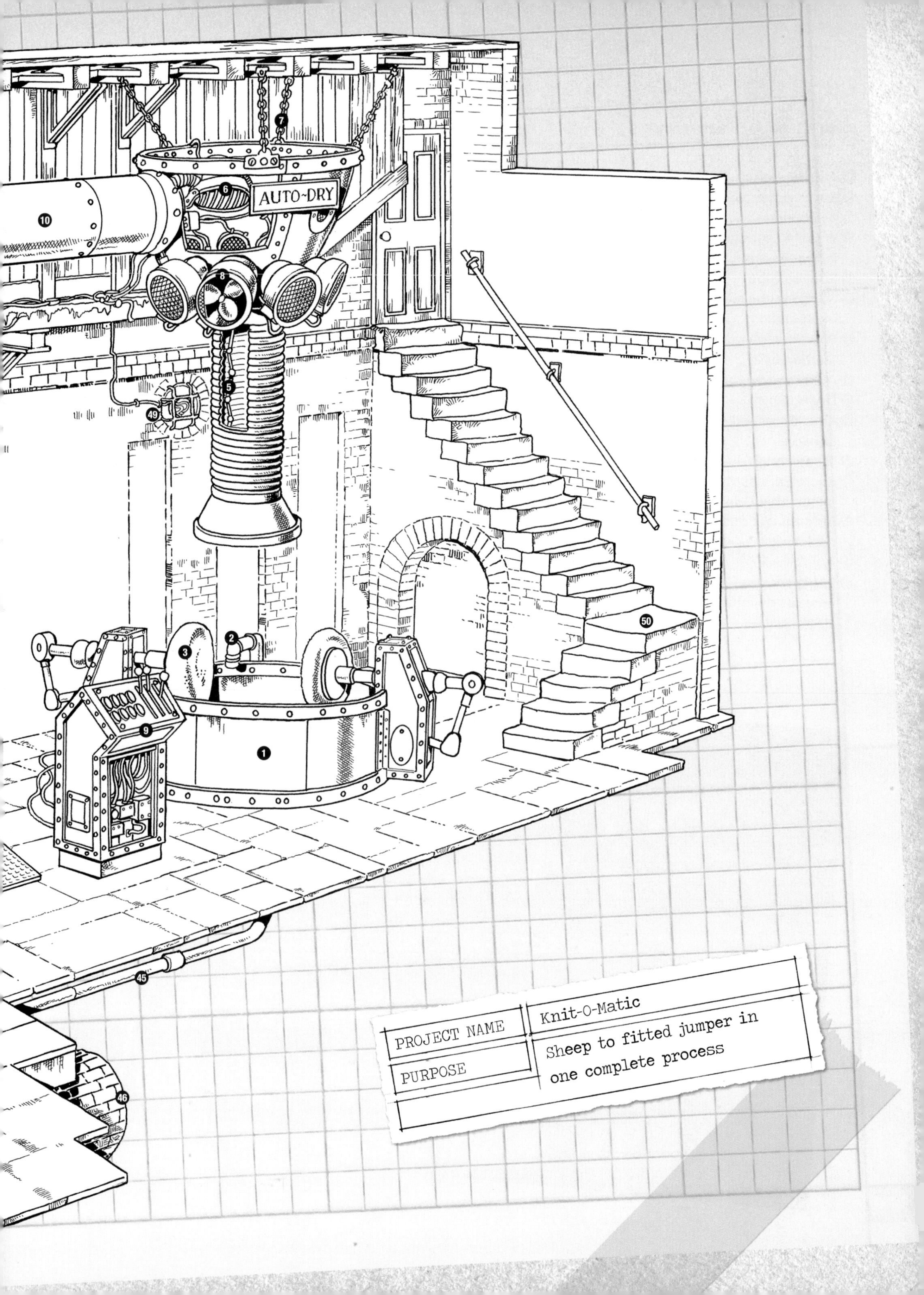
AUTO~DRY
PROJECT NAME
Knit-O-Matic
PURPOSE
Sheep to fitted jumper in one complete process

Knit-O-Matic cutaway

1. Auto Wash tub
2. Fresh water pipe (flow controlled by console)
3. Washing sponges mounted on hydraulic arms
4. Suction tube
5. Suction tube lowering mechanics
6. Powerful fan draws sheep up suction tube
7. Auto Dry support chains
8. Drying fans
9. Auto Wash and Auto Dry control console
10. Clean sheep duct
11. Secondary suction fan draws sheep along clean sheep duct
12. Auto Shaver duct (gravity assisted)
13. Auto Shaver
14. Impact reducer (trampoline) ensures no injury to sheep
15. Shearing razors
16. Hydraulic shearing arm (six in total)
17. Shearing arm electric motors
18. Variable-cut shaving control dial
19. Suction fans gather wool during shearing
20. Shorn sheep exit duct
21. Shorn sheep exit hatch
22. Wool-processing drum
23. Cord spool gearing
24. Refillable wool dye containers
25. Wool dye application tubes
26. Bucket to catch excess dye spillage
27. Wool dye refil cap and level indicator
28. Cord spool
29. Wool spinning cords
30. Flywheel
31. Auto-spinning wheel
32. Auto-knit machine
33. Yarn cone winder
34. Knitting needles
35. Knitting carriage
36. Tension dial
37. Carriage slide rail
38. Flow combs
39. Knitted jumper
40. Auto Dresser
41. Main system control console
42. Console system wiring
43. System fault indicator
44. Operator's chair
45. Water pipe
46. Drainage system and sewer
47. Drain cover allows water to drain into sewer to prevent cellar flooding
48. Wallace's Knit-O-Matic plans
49. Wiring junction box leading to main house wiring
50. Stairs to ground floor of house

size control needs adjustment

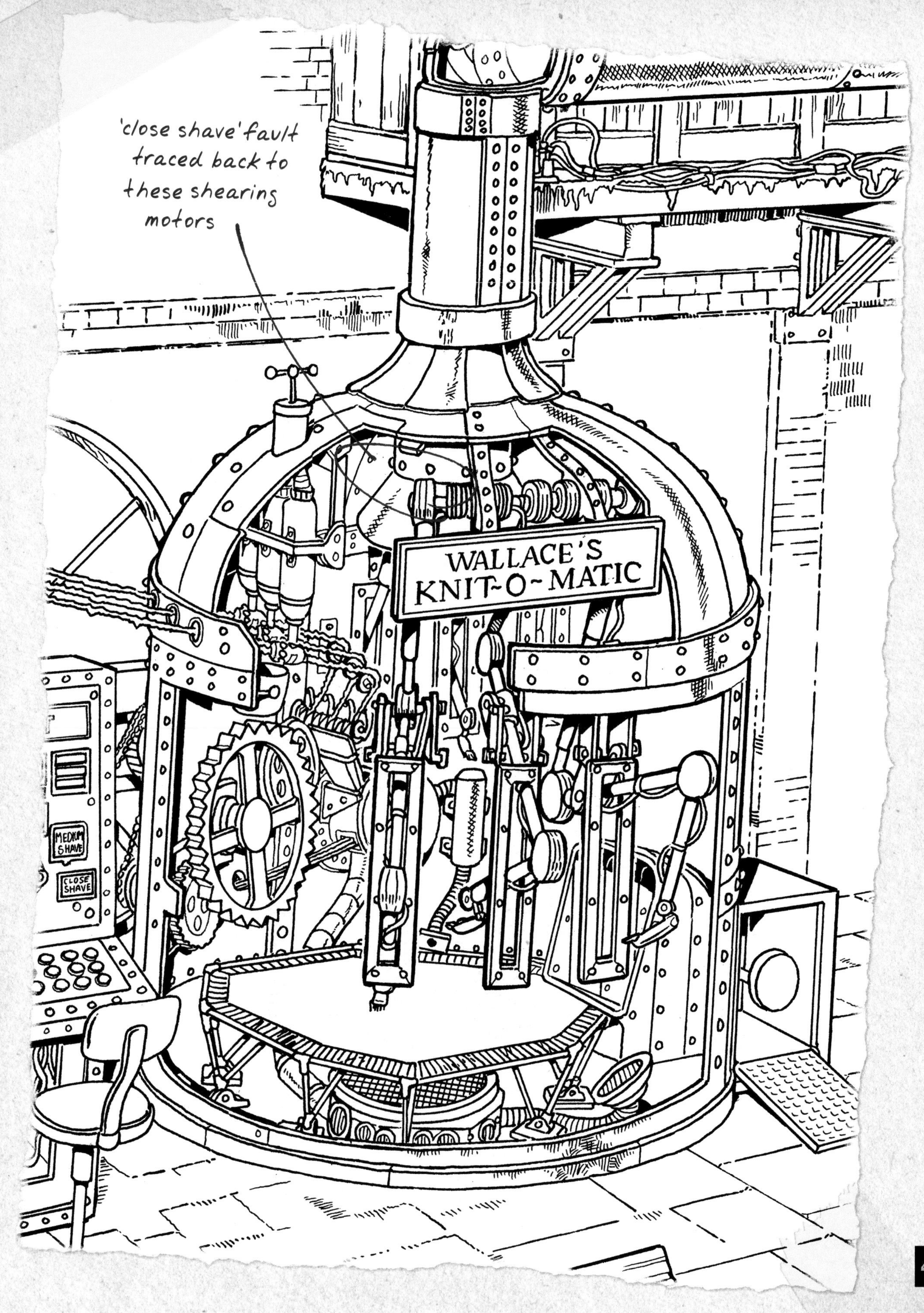
'close shave' fault traced back to these shearing motors
WALLACE'S KNIT-O-MATIC
MEDIUM SHAVE
CLOSE SHAVE

~~THE~~ MUTTON-
~~KNIT~~-O-MATIC

Contents

Now steady on!
This is a bit unsavoury. W

General description

Preston steals the plans to Wallace's Knit-O-Matic and decides to build his own version, but he adds a gruesome final stage to turn the sheep into dog food. Fortunately, Preston's plans are foiled through the heroic actions of Gromit and no living creatures are harmed by his new contraption, the ghastly Mutton-O-Matic. Even so, the machine itself is an accomplished piece of engineering and worthy of a technical description.

Freshly shorn by ~~the~~ Wallace's Knit-O-Matic, the sheep are shepherded into a holding pen. When the Mutton-O-Matic is activated, the pen is lifted by a hydraulic lifting ram, tipping the sheep on to a conveyor belt. The conveyor leads to two giant masher drums, which are fitted with hardened steel spikes. This is known as the 'primary stage' of the production process. The secondary stage involves a pair of piston-operated crushers, which compress the dog food and force it through the secondary stage filter. From there, it is sucked along a final duct and down a chute into dog food tins, which are on a moving conveyor.

Empty tins enter the machine and receive pre-glued 'Preston's Dog Food' labels from a heated applicator. After filling, the tins are fitted with lids before being packed into a waiting crate, ready for dispatch.

The entire machine is powered by a large single-cylinder diesel engine running at a steady 200rpm. This drives the primary stage masher drums and secondary stage crusher pistons, along with the reciprocating pump for the final stage suction duct and both conveyor mechanisms.

I've got patent pending on that! W

PRESTON'S
UTTON-O-MATIC

~~The~~ Mutton-O-Matic cutaway

1. Sheep holding pen (in elevated position)
2. Hydraulic lifting ram
3. Conveyor belt
4. Conveyor belt rollers
5. Conveyor belt start-stop button
6. Primary stage masher drum
7. Primary stage duct
8. Secondary stage crusher foot
9. Secondary stage crusher piston
10. Cantilever beam
11. Cantilever fulcrum
12. Secondary stage filter
13. Waste chute
14. 80 horsepower diesel engine
15. Flywheel
16. Reciprocating suction pump
17. Final stage (dog food) suction duct
18. Tin filling chute
19. Dog food tin labels
20. Heated label applicator
21. Tin conveyor rollers
22. Empty dog food tins
23. Conveyor rollers pass tins under filling chute
24. Tin lid applicator
25. Packing crate

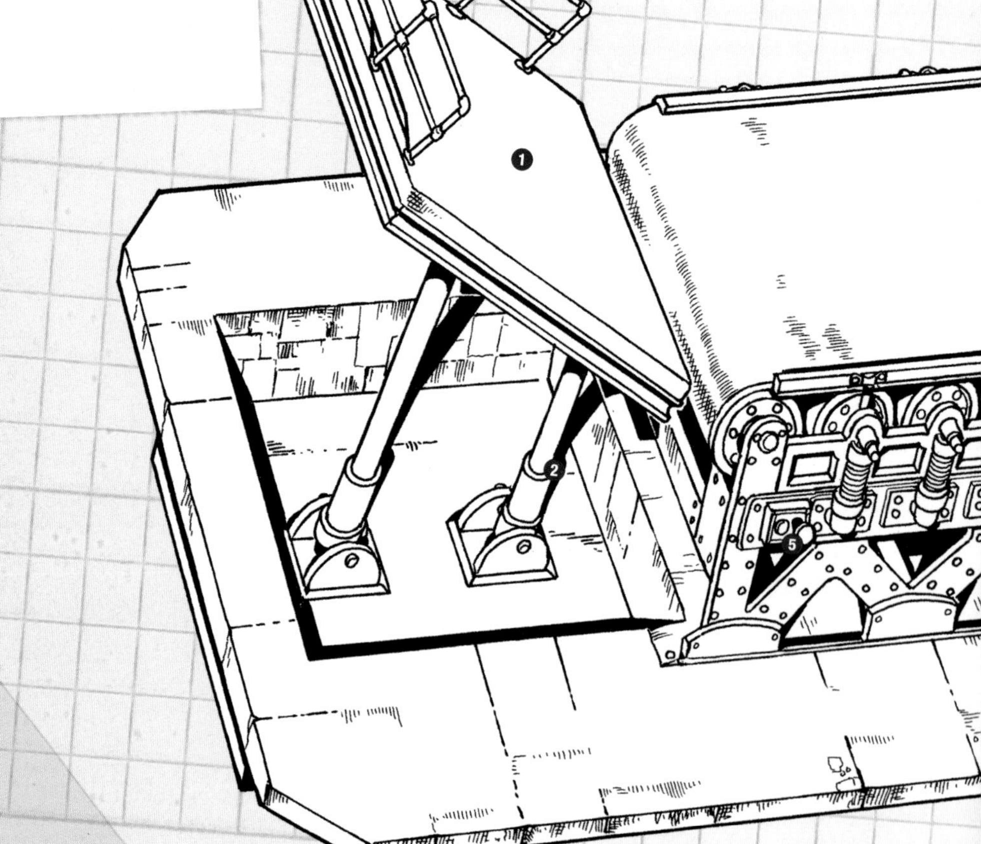

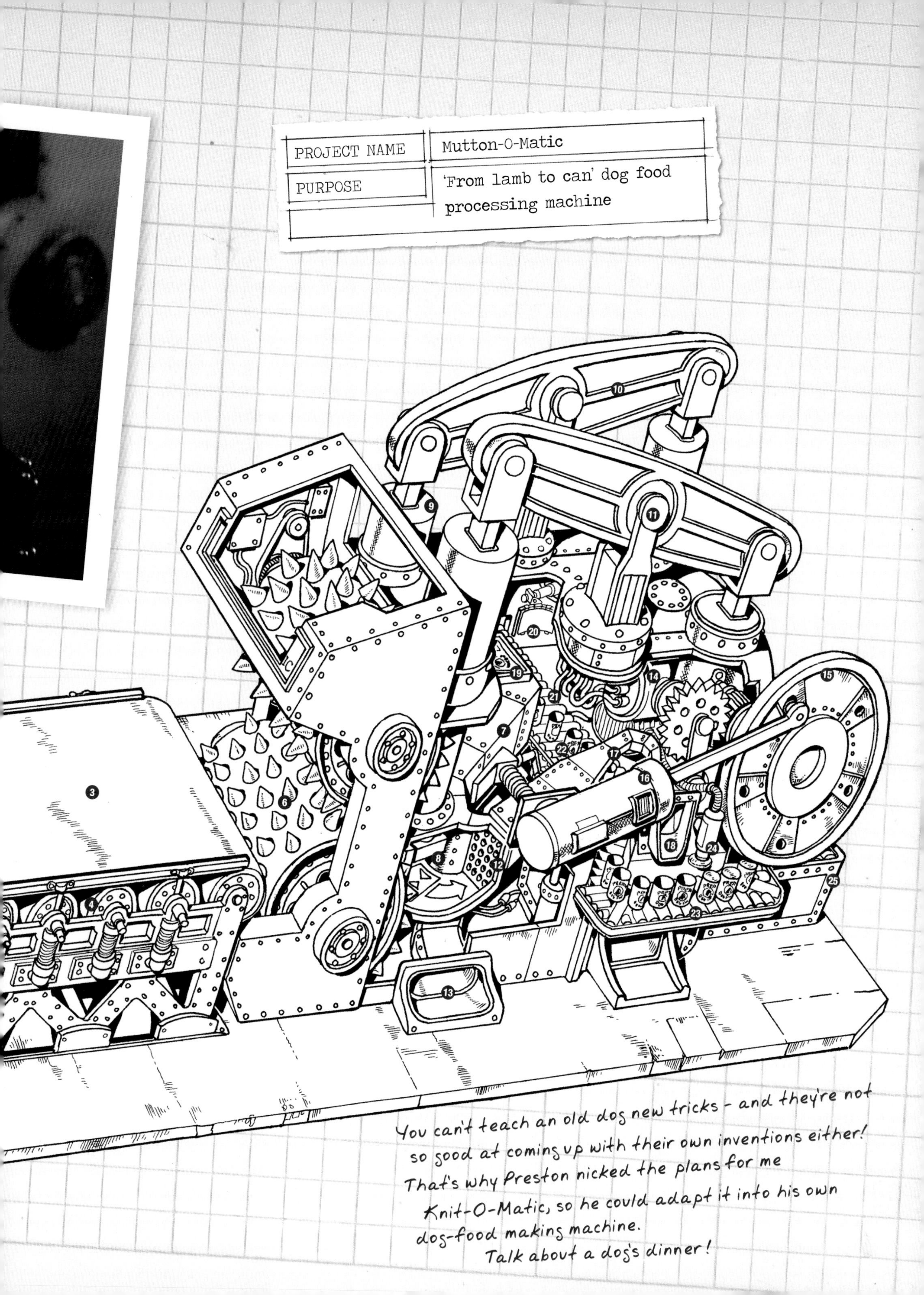

PROJECT NAME
Mutton-O-Matic
PURPOSE
'From lamb to can' dog food processing machine
3
4
6
7
8
9
10
11
12
13
14
15
16
17
18
19
20
21
22
23
24
25
You can't teach an old dog new tricks - and they're not so good at coming up with their own inventions either!
That's why Preston nicked the plans for me Knit-O-Matic, so he could adapt it into his own dog-food making machine.
Talk about a dog's dinner!

PRESTON THE CYBER DOG

Contents

General description

When Wendolene Ramsbottom's father died he left her the family wool shop along with "debts and a few other things", namely Preston, a sophisticated cyber dog which he built to protect both his daughter and the business. Unfortunately, Preston carries out his programming with ruthless efficiency, even resorting to sheep rustling in order to supply the shop with wool.

Preston's body is built around a skeleton of high-strength titanium alloy. The skeleton, skull and internal mechanics and electronics are covered with artificial skin and fur so that, from the outside, Preston appears to be a normal dog. He is equipped with two stereoscopic cameras (eyes), two articulated unidirectional microphones (ears) and a multi-spectrum odour detection sensor (nose). He is also able to 'eat' like a real dog, with food being sucked down two tubes leading from his mouth to a small incinerator, located just below the neck.

All of Preston's movement is driven by a powerful electric motor, positioned centrally within the torso. Individual leg, paw and jaw movements are controlled via a series of pulleys, gears and cranks. A gyroscope fitted at the bottom of the torso works together with the unidirectional microphones to help maintain balance, especially when Preston is walking upright.

The main computer, housed in the chest area, provides overall control of Preston's functions, while memory and programming data is stored on magnetic tape.

Preston operates independently, and he can learn and adapt in order to protect himself, but he has been programmed with protocols to ensure that he never harms Wendolene or the family business. However, over time Preston's increased self-awareness begins to override his programming, with dramatic consequences.

After framing Gromit for the sheep rustling, Preston steals the plans for Wallace's Knit-O-Matic machine and builds his own version. However, having decided to go into the dog food business, he adds an extra stage of his own design: the Mutton-O-Matic (see page 46). This horrendous invention is a fully functioning prototype, and Preston's first victims are the sheep, including Shaun, together with Wallace and Wendolene.

Preston the Cyber Dog cutaway

❶ While disguised as a real dog, food tubes suck food to a small incinerator for disposal
❷ Incinerator located behind electric motor
❸ Incinerator vacuum pump powered by adjacent electric motor
❹ Main power switch (locked in 'On' position)
❺ Unidirectional microphones provide spatial awareness and help to maintain balance
❻ Independently suspended cameras provide stereoscopic vision
❼ Visual processing cortex
❽ Multi-spectrum ultra-sensitive odour detection system
❾ Power and data conduits connect audio and visual processors to computer
❿ Electric motor
⓫ Heat sink tubes built into collar spikes dissipate heat from motor
⓬ Jaw hinge mechanism
⓭ Paw and claw grip control actuators
⓮ Titanium ultra-grip claws
⓯ Left fore leg and paw control gearing
⓰ Torso movement hydraulic jack
⓱ Left hind leg control gearing
⓲ Multi-jointed hind legs support Preston's weight and provide most of the motive power
⓳ Claws provide additional grip and balancing adjustments
⓴ Right hind leg control gearing
㉑ Gyroscope maintains robot's balance in upright position on hind legs
㉒ Flexible torso joint seal
㉓ Batteries
㉔ Battery support stanchion
㉕ Main computer and memory bank, programmed with self-awareness protocols that cannot be overridden
㉖ Memory tape spool controls (locked to 'On' position and tamper-proof)
㉗ Memory/programming tape spools
㉘ Illuminated memory code programming controls

The very last photo of the original Preston

PROJECT NAME	Preston the Cyber Dog
PURPOSE	Protection of Wendolene and the family wool business

A dog is a man's best
friend apparently.
Well, blow me down if
Preston didn't turn out
to be a robot dog - and not
a very friendly one at that!
1
2
3
4
5
6
7
8
9
10
11
12
13
14
15
16
17
18
19
20
21
22
23
24
25
26
27
28

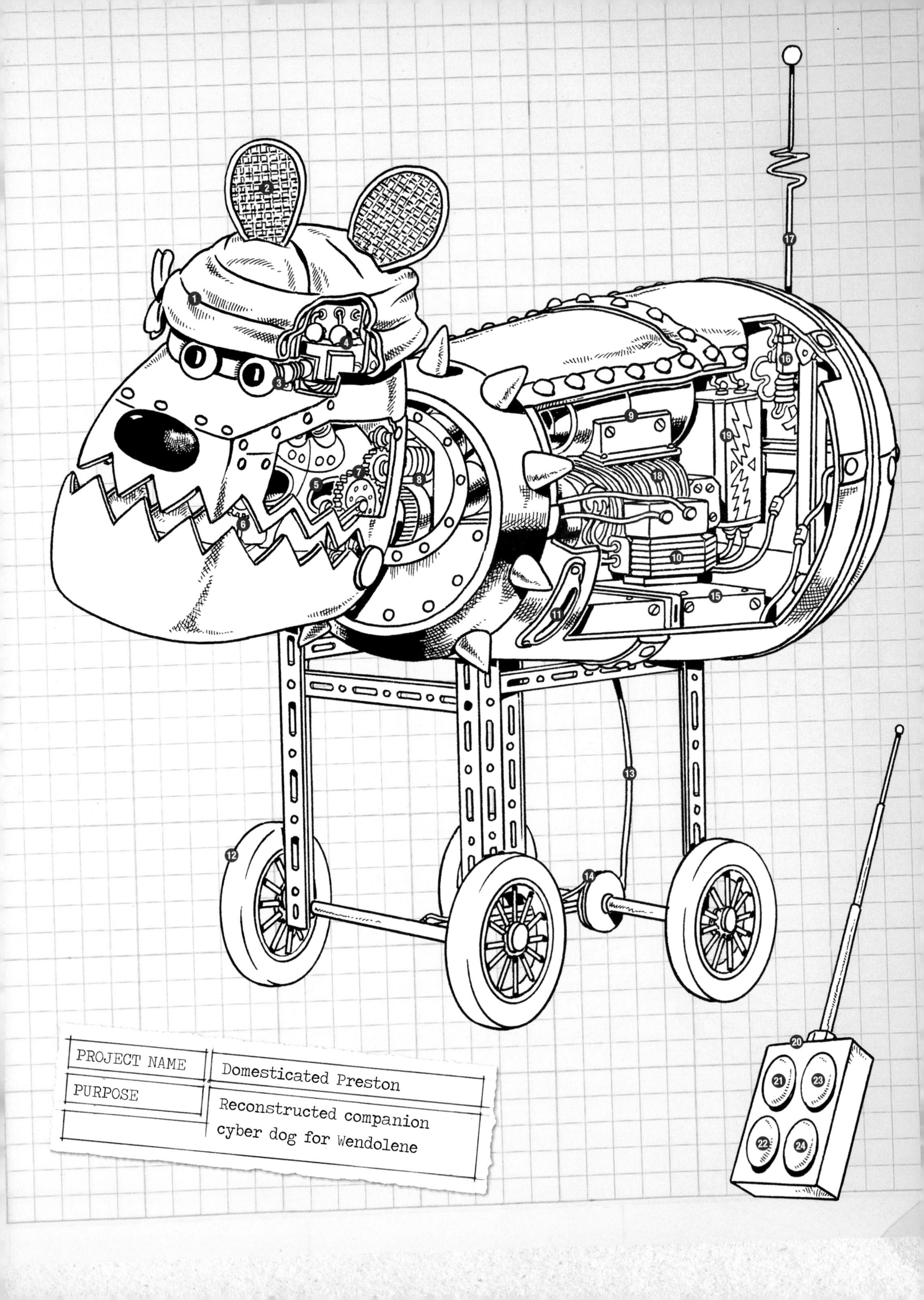
PROJECT NAME
Domesticated Preston
PURPOSE
Reconstructed companion cyber dog for Wendolene

Following the dramatic events involving Preston and his Mutton-O-Matic machine, Wallace salvages what pieces he can find of the old Preston and uses them to build Wendolene a more friendly 'domesticated' cyber dog. A new body is constructed of riveted steel plate and attached to a simple trolley frame and wheels, allowing the new Preston to move around. The head is reshaped and the jaw fitted with an anti-bite mechanism, while the stereoscopic cameras and unidirectional microphones have been reused. A repair bandage around Preston's head also provides additional thermal insulation for the damaged circuitry inside. Most importantly, Preston has been reprogrammed with restricted self-awareness protocols, and he is now fully controlled by a hand-held remote unit which features a dedicated pre-programming override button, just in case...

Domesticated Preston cutaway

1. **Repair bandage and thermal insulation**
2. **Unidirectional microphones**
3. **Reconditioned stereoscopic cameras**
4. **Visual processing cortex**
5. **Dog food suction pipes (redundant)**
6. **Anti-bite clamp**
7. **Jaw hinge gearing**
8. **Power and data conduits connect audio and visual processors to main computer**
9. **Incinerator (redundant)**
10. **Incinerator vacuum pump (redundant)**
11. **Original fore leg joint socket**
12. **Fore and hind legs replaced with easy-to-control trolley frame**
13. **Rear-wheel drive power cord**
14. **Speed and braking control regulator**
15. **Reprogrammed control computer, now with restricted self-awareness protocols, which can be overridden by remote control**
16. **Remote-control transceiver**
17. **Remote-control antenna**
18. **Electric motor**
19. **Battery**
20. **Remote-control unit**
21. **Forward/backward control**
22. **Pre-programming override**
23. **Steering control: once for left; twice for right**
24. **Start/stop button**

Not a bad job, if I may say so

MOTORCYCLE & SIDECAR

Contents

General description

Wallace's motorcycle is based on a traditional British design by Triumph although it will be clear to the enthusiast that various modifications have been made over the years. Having said that, these are all fairly minor and appear to have no significant effect on its operation. The motorcycle is powered by a 350cc single-cylinder four-stroke engine, which drives the rear wheel via a four-speed manual gearbox and chain. Braking is by cable-operated drum brakes front and rear. Otherwise the motorcycle is conventional in nearly every respect.

Of far greater interest is the sidecar, which has undergone a number of significant modifications and contains two special features that are worth noting in particular. The first of these is its ability to transform into an aeroplane. This requires the sidecar to be separated from the motorcycle, before panels in the side and rear of the aluminium bodywork open to allow wings and a tailplane to be extended. In the nose of the sidecar, slightly offset to the right, is a compact 'aeromatic' petrol engine, which drives a propeller via a toothed belt. The propeller itself is stored with its blades folded within the nose of the sidecar but is deployed instantly at the same time as the wings. The transformation from sidecar to aeroplane is activated by buttons on a control panel in front of the passenger/pilot, and the whole process takes less than a second – which proves to be not a moment too soon when the sidecar separates from the motorcycle by accident and Gromit finds himself plunging to the bottom of a 2,000ft drop.

Located behind the engine and propeller assembly is the sidecar's second special feature: the sud cannon. This is a modified version of the porridge cannon found at 62 West Wallaby Street and is used by Wallace and Gromit for their window-cleaning business. Dried detergent and water are stored in separate tanks and mixed together inside the sud cannon before being drawn into the cannon barrel. Compacted shots of high-density soap foam are then fired from the sidecar at ground level (usually by Wallace) while Gromit climbs the ladder to clean the windows. All other aspects of the sud cannon's operation are very similar to that of the porridge cannon.* In fact, during his attempts to stop Preston from rustling sheep, Gromit fills the sud cannon's detergent tank with extra-thick porridge mix.

* see page 36

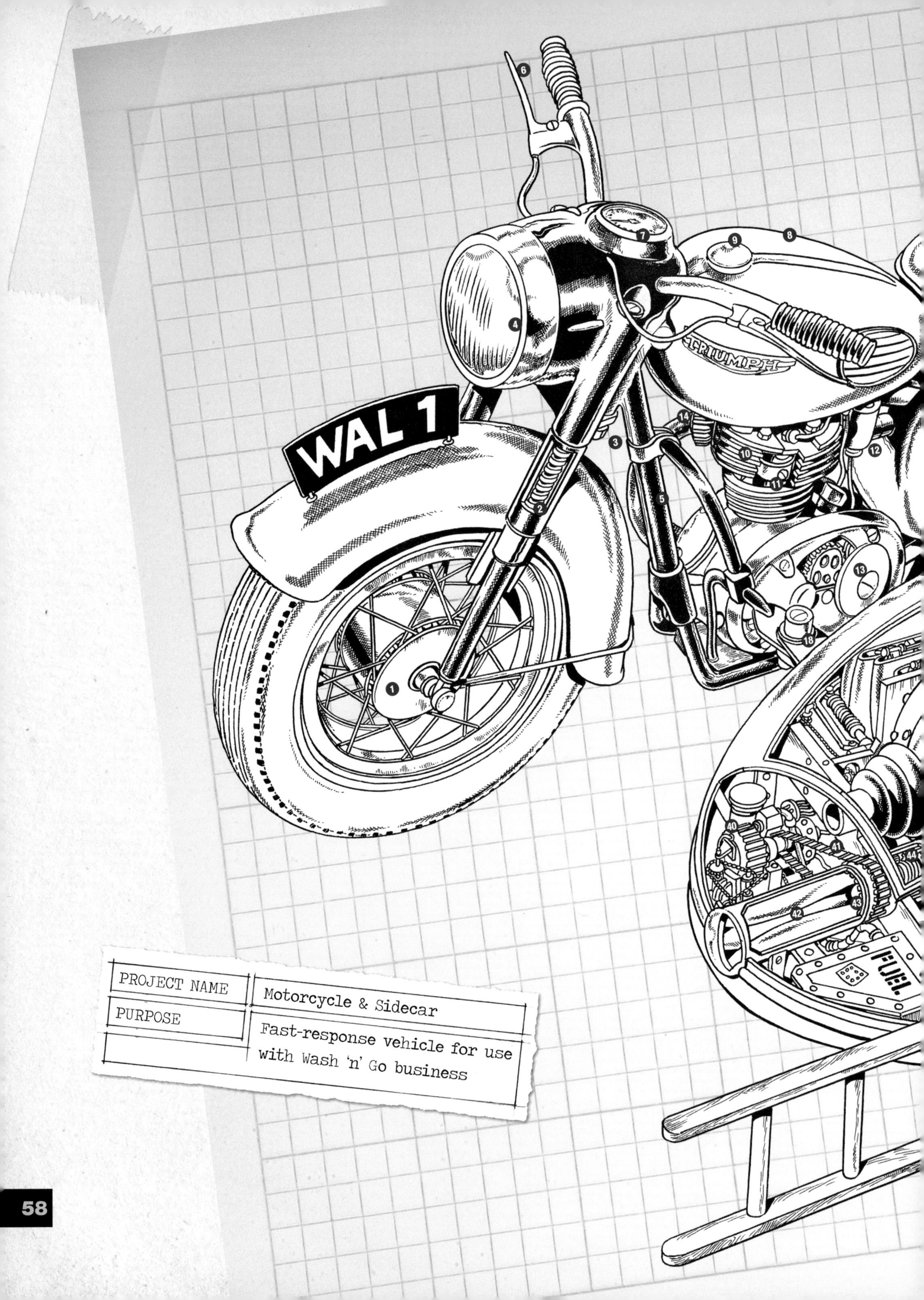

WAL 1
TRIUMPH
FUEL
PROJECT NAME
Motorcycle & Sidecar
PURPOSE
Fast-response vehicle for use with Wash 'n' Go business

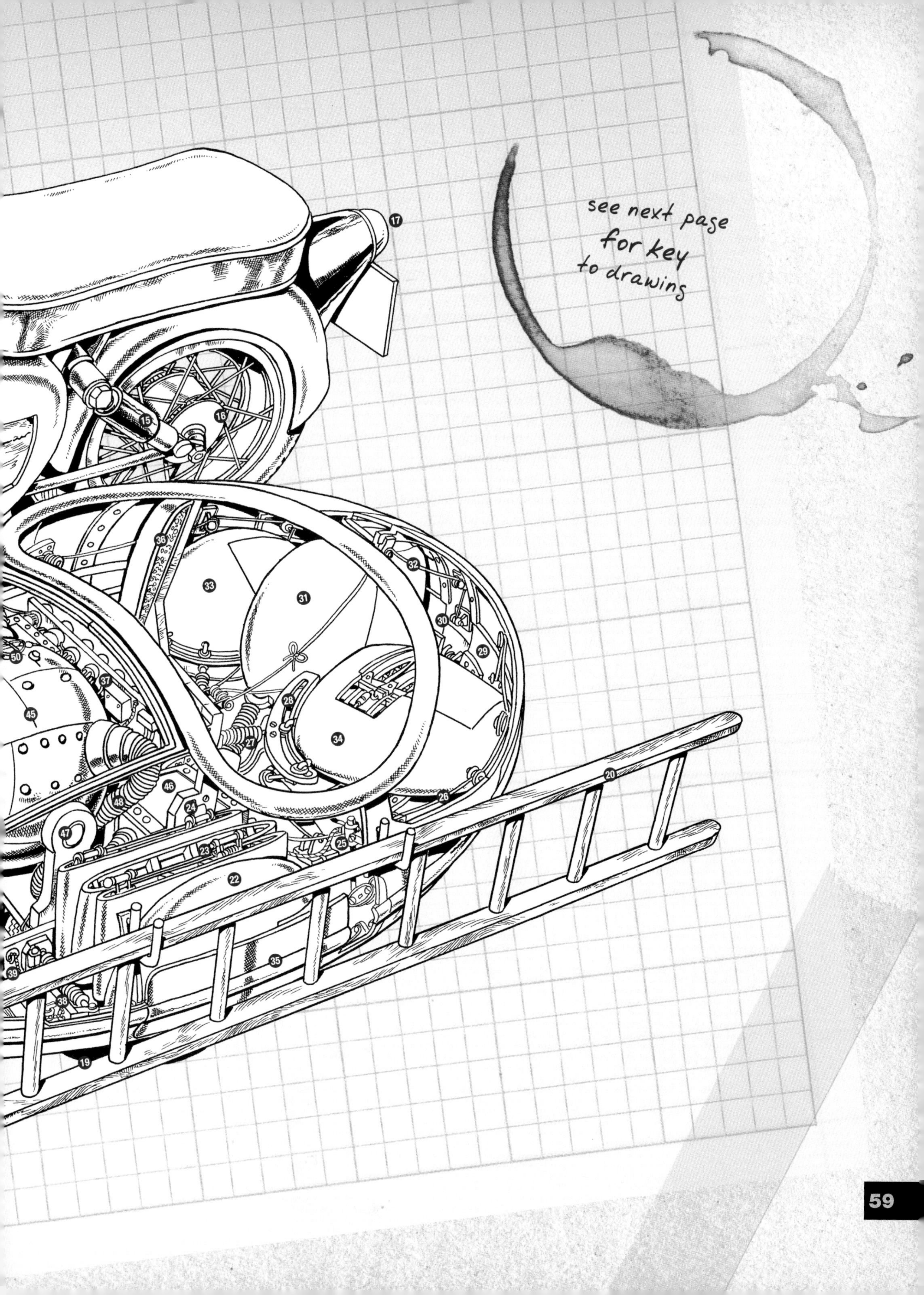
see next page
for key
to drawings
17
15
16
36
32
33
31
30
29
50
37
45
28
27
34
20
46
48
24
26
47
23
25
22
39
35
38
19

Motorcycle & Sidecar cutaway

1. Front brake drum
2. Hydraulically damped front forks
3. Steering head
4. Headlamp
5. All-welded tubular motorcycle frame
6. Front brake lever
7. Speedometer
8. Fuel tank
9. Fuel cap
10. Single-cylinder four-stroke engine
11. Piston
12. Carburettor
13. Gearbox
14. Exhaust pipe
15. Rear suspension damper
16. Rear brake
17. Rear number plate
18. Front sidecar securing bolt
19. Sidecar wheel
20. High-window access device
21. Retracted starboard wing
22. Retracted port wing (folded in three sections)
23. Wing fold operating link
24. Wing spar attachment bolt
25. Port wing panel actuator
26. Wing panel control string
27. Wing deployment spring
28. Wing positioning slide
29. Rear port wing panel
30. Tailplane panel
31. Tailplane
32. Tailplane panel control strings
33. Starboard rear wing
34. Port rear wing
35. Port wing panel
36. Sidecar/pilot's seat
37. Flight controls
38. Port wing deployment spring
39. Wing deployment control string
40. Aeromatic engine
41. Propeller drive belt
42. Three-blade propeller (shown retracted)
43. Spring-loaded propeller hub
44. Propeller hub spring
45. Window-cleaning detergent cannon
46. Dried detergent mix tank
47. Water tank
48. Detergent tank to mixer pipe
49. Cannon deployment hydraulics
50. Fold-down target finder

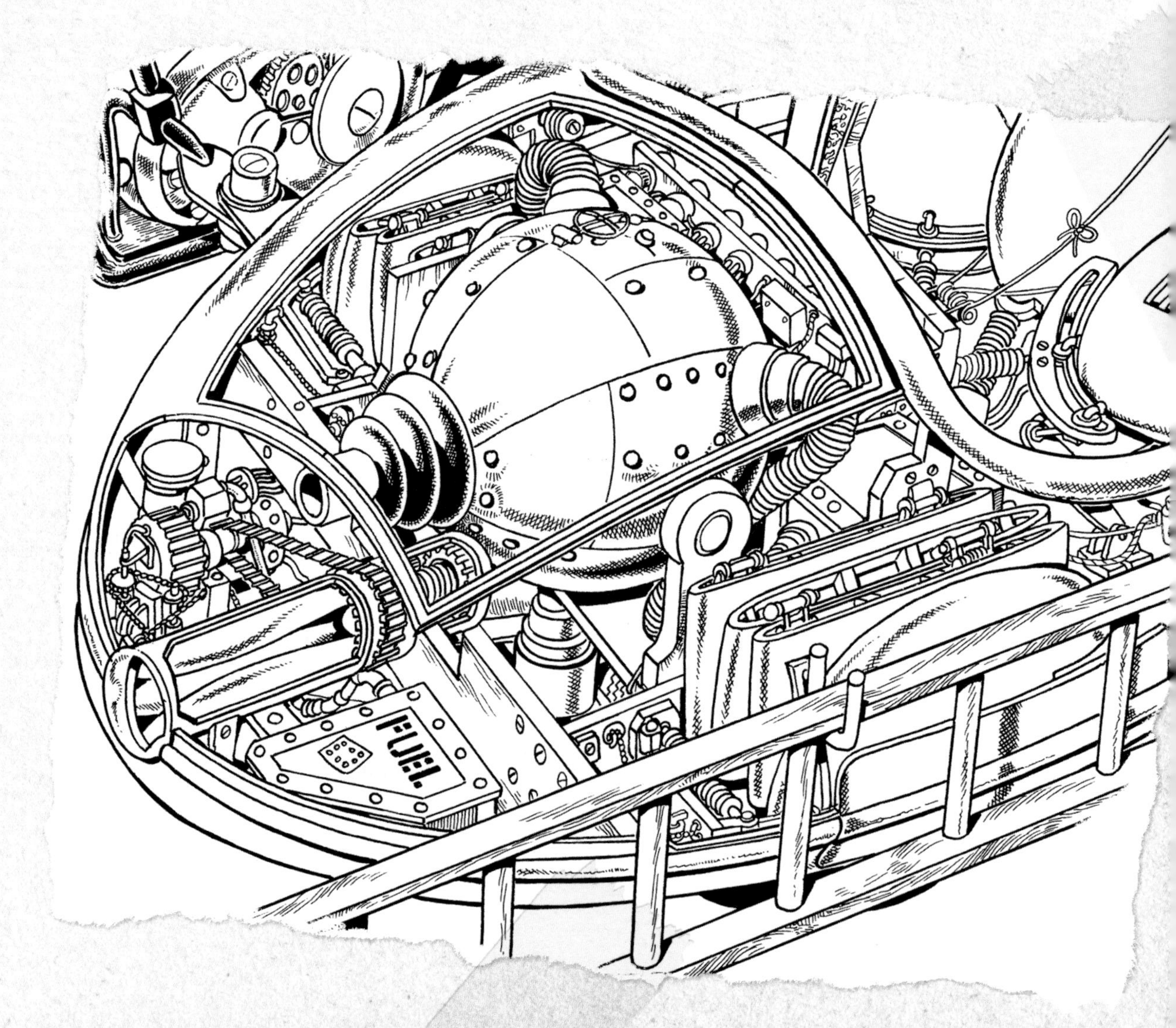

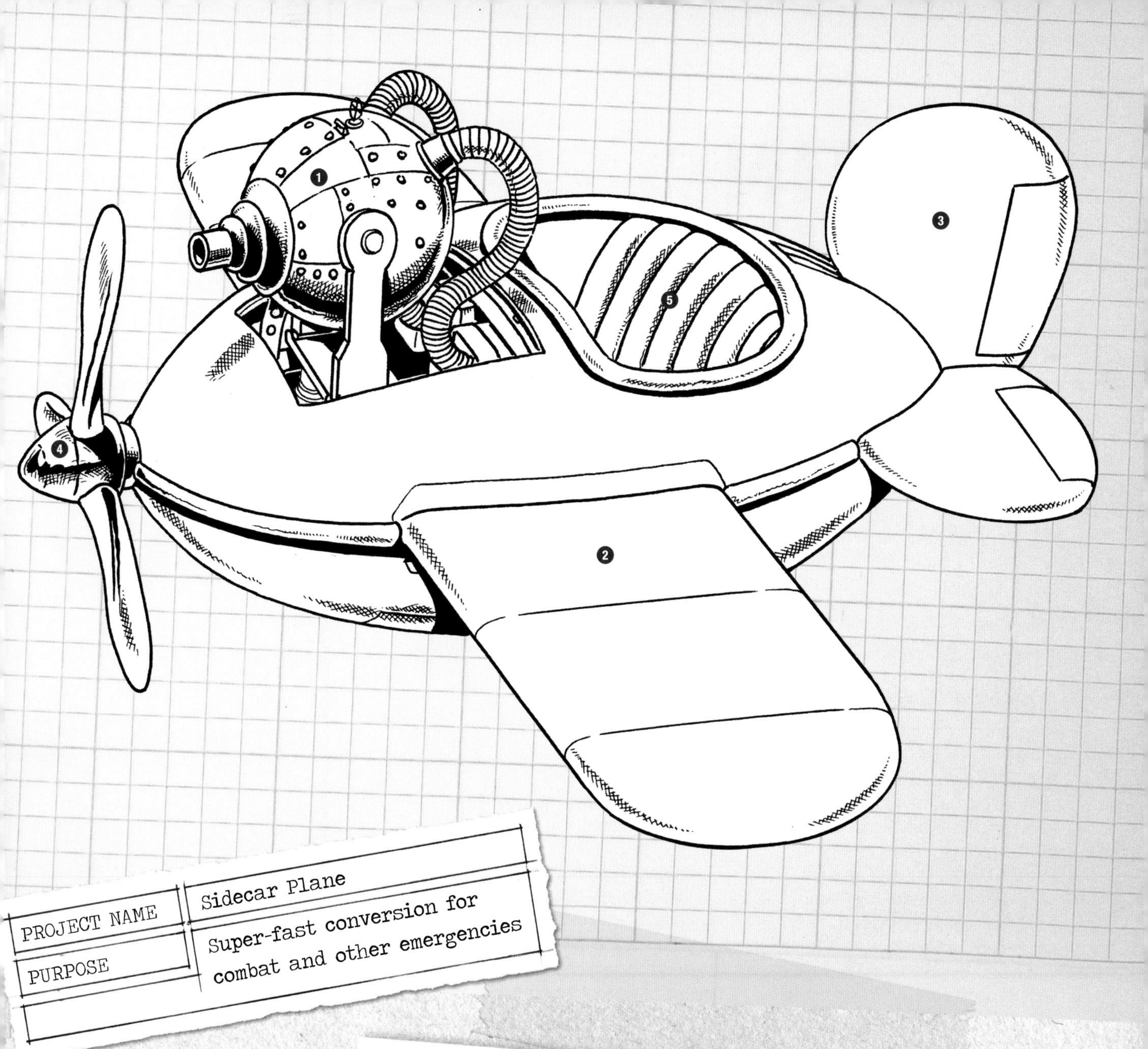

PROJECT NAME	Sidecar Plane
PURPOSE	Super-fast conversion for combat and other emergencies

1. Sud cannon (deployed)
2. Port wing
3. Tailplane
4. Propeller
5. Pilot's seat

AUSTIN A35 VAN

Contents

General description

Wallace and Gromit are great fans of the Austin A35 Van and throughout their many adventures and various money-making enterprises they have used at least two such vehicles to provide reliable transport along with plenty of space to carry cargo and contraptions. Their latest van was built in 1965 and is still fitted with its original 1098cc, four-cylinder A-series engine and gearbox. It is shown here in the livery of the Top Bun bakery business, which Wallace and Gromit run from their home at 62 West Wallaby Street.

Although from the outside the van appears to be unmodified, it does, in fact, have a few special features. The most significant of these are the seats, which can be installed and removed via access hatches in the floor pan. The seats themselves are normally stored 'on standby' in the cellar below the garage. When it's time to make the busy morning delivery round, every second counts and the seats, along with the driver and passenger, are raised swiftly up through the garage floor and then through the floor of the van on hydraulic lifting rams. Once inside the cab, the clamps securing the seats to the hydraulic rams are released, and the rams withdraw back to the cellar while the hatches in the floor pan of the van are closed and the seats anchored in place.

In the cab, the steering wheel is normally positioned on the right-hand side but can be removed from the steering column and attached to a duplicate steering column on the left-hand side for continental driving. The swap-over can be achieved very quickly in the event of an emergency, where the driver must ask the passenger to "take the wheel". Beneath the radio unit in the centre of the dashboard is a slot-loading long-play (LP) record player, driven directly by the engine. This is a recent addition to the vehicle and was installed to replace the Auto-Burn Ⓟ toaster fitted previously. A selection of LPs is stored in a record case built into the lower dashboard on the passenger side.

The rear of the van is fitted with a bread shelf transit frame, which is designed to carry removable bread shelves. These are loaded with bread fresh from the ovens at the Top Bun bakery, ready for dispatch early each morning.

DOH NUT5

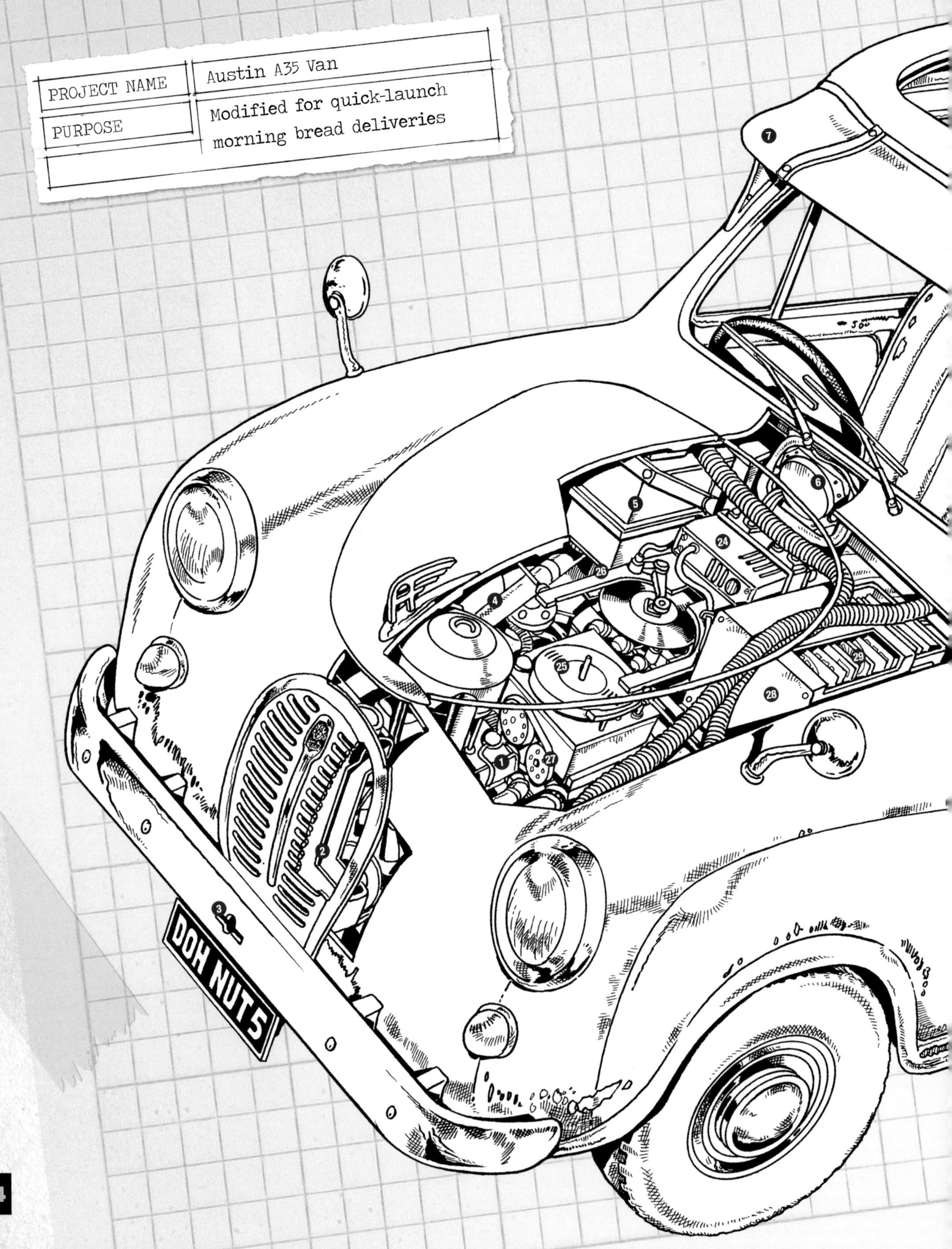
PROJECT NAME
Austin A35 Van
PURPOSE
Modified for quick-launch morning bread deliveries
DOH NUT5

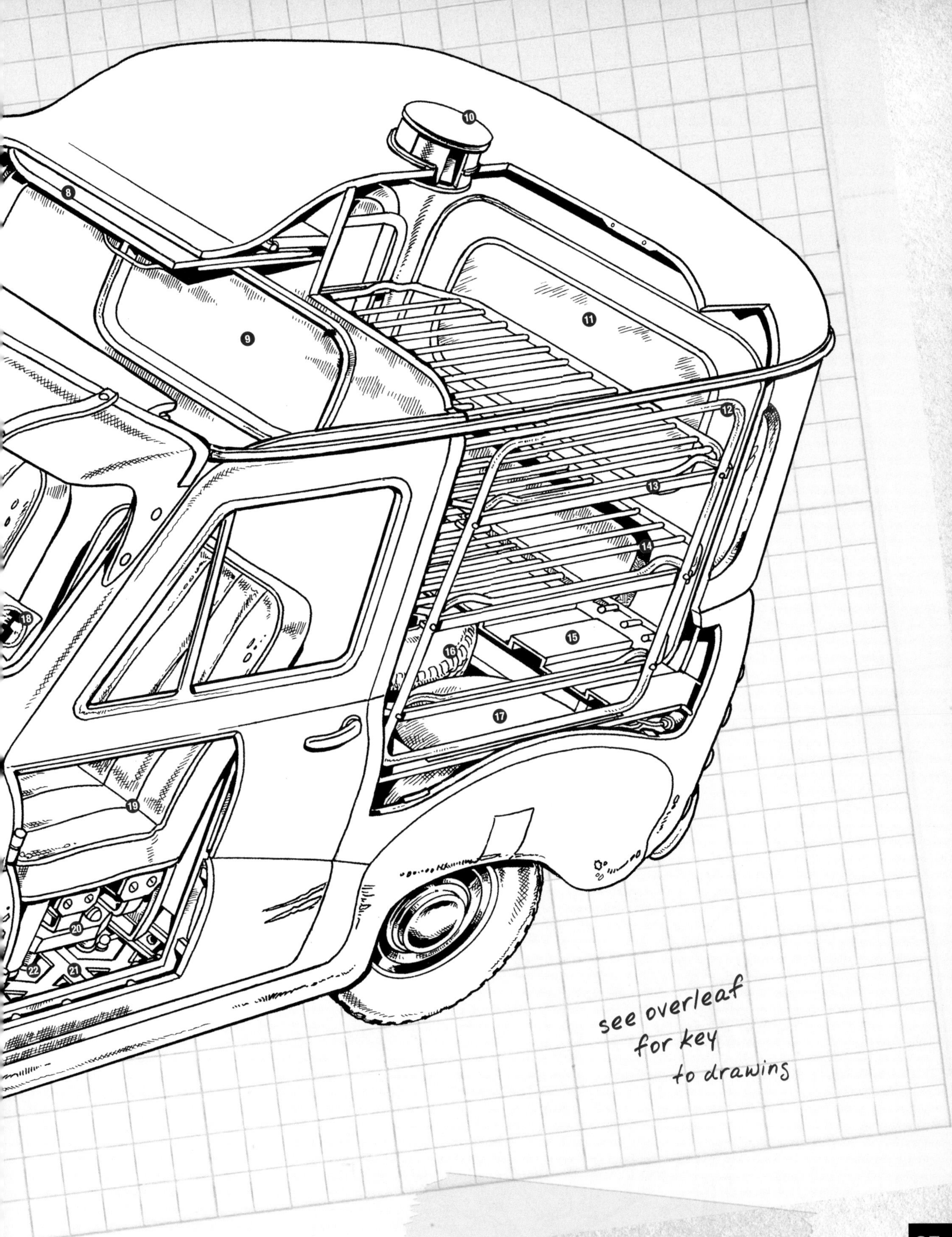
8
9
10
11
12
13
14
15
16
17
18
19
20
21
22
see overleaf
for key
to drawings

Austin A35 Van cutaway

1. Engine
2. Radiator
3. Manual crank handle slot
4. Right-hand drive steering column (duplicated on left for continental driving)
5. Battery
6. Speedometer
7. Sun visor
8. Sun roof (retracted to allow Wallace and Gromit to wear bakers' hats while delivering bread)
9. Rear of cab 'bread-checking' window
10. Ventilation fan
11. Rear loading door
12. Bread shelf transit frame
13. Transit frame shelf slide runners
14. Adjustable and removable bread shelves
15. Rear load floor
16. Spare wheel
17. Inner wheel arch
18. Up-to-date tax disc
19. Passenger seat (normally stored on hydraulic lift in cellar below garage of 62 West Wallaby Street)
20. Strengthened passenger seat subframe
21. Under-floor access hatch through which passenger enters van
22. Passenger seat hydraulic lift clamps
23. Gear lever
24. Radio
25. Single-speed vibration-damped auto-return record player
26. Record player LP slot-loader (this was added to replace the Auto-Burn toaster Ⓟ located under radio)
27. Record player driven directly from engine.
28. Aluminium record-storage cabinet
29. Cracking collection of long-playing records

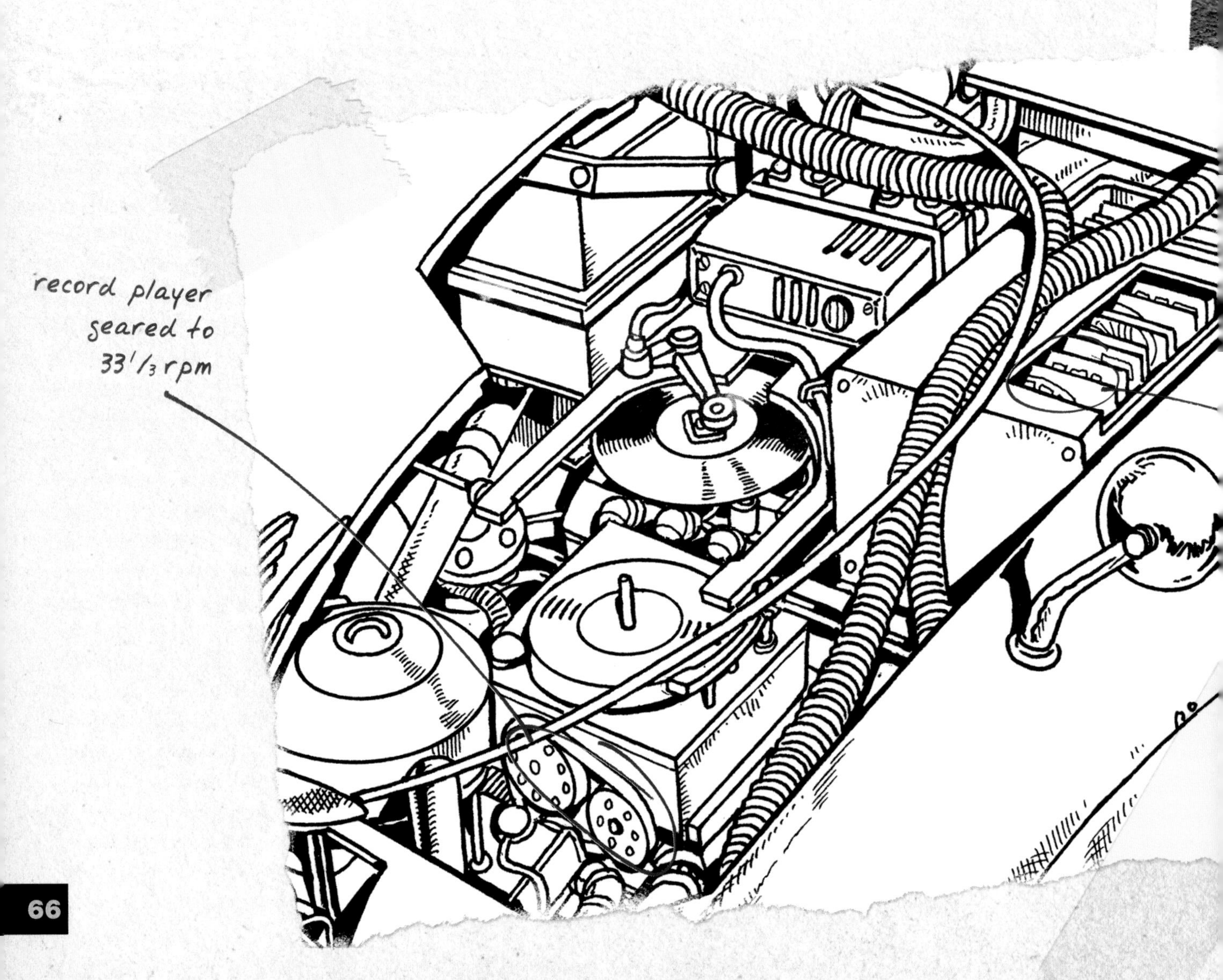

Gromit's
record
collection

TOP BUN BAKERY

Contents

General description

When Wallace and Gromit start their Top Bun bakery business, they carry out extensive modifications to the house at 62 West Wallaby Street. The most obvious of these is the traditional cloth-sail windmill, which is installed on the roof. This provides on-site milling of grain to make flour as well as power for the automated dough kneading machine, mixer, bread ovens and other bakery equipment.

The fast-response launch system, originally installed for Wallace and Gromit's Wash 'n' Go window-cleaning business, has been updated for use with the Top Bun delivery van (see Austin A35 Van, page 62).

The early-morning bread delivery round meant that a new 'GET-U-UP' system had to be devised to get Wallace out of bed at 5am and ready to leave as quickly as possible. A modified Bed Launcher now delivers Wallace directly to the delivery van, while on the way he is automatically dressed in his clothes and baker's hat. Once in the van, Wallace has his tea mug filled from a pump nozzle, which can also supply 4-star petrol, diesel, coffee and milk. The system is neat and efficient although the tea can be prone to a 'diesely aftertaste'.

Most of the various other contraptions found in the Top Bun bakery are dedicated to the bread-making process, which is mostly automated. However, the forklift truck is one exception.

Although largely conventional in design and operation, the 'arms' of the forklift truck have been specially modified to include two highly articulated mechanical grabber hands. These are controlled by a pair of multi-position servos fitted either side of the mast. The grabber hands are able to grasp hot bread trays from the oven and load them directly into the back of the delivery van. Specially designed heat-resistant oven gloves are fitted over the grabber hands to protect them and to provide additional grip.

The forklift truck is powered by a petrol engine, which is fuelled by liquified petroleum gas (LPG) from a bottle mounted behind the operator. This can be replaced when empty to allow continuous operation since there is no need to wait while batteries are recharged. The vehicle is driven by the front wheels and steered by the rear wheels, this arrangement offering a good combination of traction, stability and manoeuvrability.

PROJECT NAME
Forklift truck
PURPOSE
Safe handling of hot bread shelves from oven to van

Forklift truck cutaway

1. Driving wheels
2. Steering wheels
3. LPG engine
4. Fuel control valve
5. LPG regulator/vaporiser
6. Filter
7. Ventilation grille
8. LPG fuel tank
9. Counterweight
10. Handbrake lever
11. Forward/reverse lever
12. Padded seat for operator comfort
13. Forklift controls
14. Forklift control gearing
15. Drive gearing
16. Steering gear
17. Accelerator pedal
18. Steering wheel
19. Mast
20. Crosshead
21. Lifting chains
22. Lifting arm hydraulic ram
23. Multi-position arm servos
24. Toughened bread tray grips
25. Heat-resistant oven gloves
26. Bread tray

AUTOCHEF

Contents

General description

The Autochef is a fully mobile, remote-controlled robot chef, designed to cook and serve breakfast at the push of a button. It is one of Wallace's ongoing projects and – it is fair to say – still in development.

While the prototype Autochef is undergoing testing, the remote control is configured with a basic 'menu', which includes eggs, served either scrambled or fried, and a choice of tea or coffee, with or without sugar. Further functions, and a more extensive menu, are planned for the future.

Internally, the Autochef can be divided into roughly two halves. The bottom half contains the wheels, drive motor and associated gearing that enable the robot to move independently between the kitchen and the dining room. At the bottom of the casing is a water tank with an integral heating element, which supplies hot water for tea and coffee. The tank is refilled via a funnel, which is accessed when the lid of the Autochef is opened.

Mounted above the water tank, on a pair of scissor lifts, is the hot drinks platter. This contains all the necessary ingredients (apart from the hot water) to prepare and serve tea and coffee. Hot water is supplied from the tank via an electric pump to the tea and coffee pots; milk and sugar can be added at this stage if required. Once ready, the platter is raised up through the body of the Autochef to the 'neck' and the drinks are served through spouts which extend from the tea and coffee pots and through a hole in the 'face'.

The top part of the casing contains a frying pan, which is hinged at one side, allowing it to swing down to a vertical position when the tea and coffee platter is in the raised position. The frying pan can also swing up in order to serve cooked food, and this requires that the lid of the Autochef (also hinged) is opened first. Below the frying pan is a hinged heating ring, which swings into a horizontal position underneath the frying pan for the cooking of eggs.

At the top of Autochef's 'head' is a food measuring and mixing jug with integrated high-speed blending blades. The 'face' contains a mixer speed dial, a multi-function timer dial and a temperature gauge, while control circuitry and wiring is mounted behind on the inside of the casing.

Operation of the Autochef is straightforward although, unfortunately, it is still somewhat unreliable.

PROJECT NAME
Autochef
PURPOSE
Remote-controlled food
cooking and serving r
DAIRY MI
TEA
TEA
COFFEE
SUGAR
5
6
7
8
9
10
11
12
13
14
16
17
18
19
20
21
22
23
24
25
26
27
28
29
30
31
32
33
34
35
36
37
38
39
40
41
42

Autochef cutaway

1. Autochef activation buttons
2. Remote-control antenna
3. Temperature slider control
4. Timer control dial
5. Food mixer bowl
6. Food mixer bowl lid
7. Food mixer speed dial
8. Multi-function timer dial
9. Temperature gauge
10. Food mixer/dials wiring
11. Frying pan (in horizontal position for normal use)
12. Frying pan hinge
13. Heating ring wiring
14. Heating ring (swings down when not in use)
15. Lid hinge allows access for maintenance and refilling
16. Autograb unscrews milk bottle lid
17. Refillable milk bottle
18. Milk pouring arm (also used to place teabags in pot)
19. Sugar deployment and stirring arm
20. Sugar/coffee spoon
21. Arm control motors
22. Remote-control antenna
23. Control systems wiring
24. Electric pump circulates hot water from tank to tea and coffee pots
25. Water tank refill funnel
26. Remote-control transceiver
27. Coffee caddy
28. Teabag caddy
29. Sugar caddy
30. Teapot
31. Telescopic spout on both tea and coffee pots
32. Telescopic spout motor
33. Hinged lids on tea and coffee pots
34. Revolving platter positions tea and coffee pots behind pouring hole
35. Platter rotation gearing
36. Platter scissor lift
37. Electric motor for platter scissor lift
38. Water tank
39. Water tank heating element
40. Flexible piping connects water tank to refill funnel and pump
41. Steering caster wheels
42. Pouring hole (also acts as a vent when Autochef is in cooking mode)

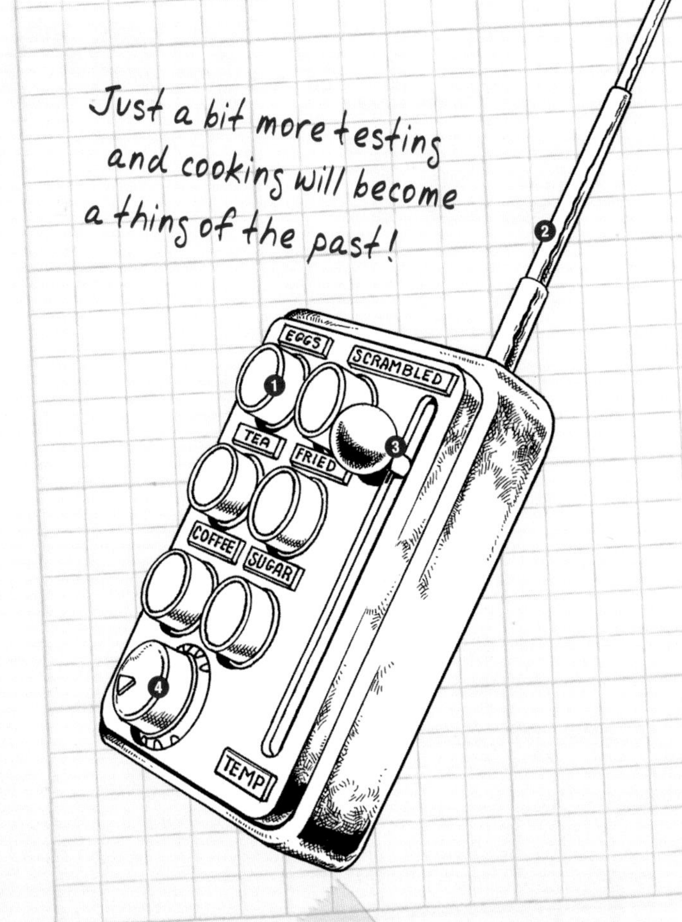

BULLY-PROOF VEST

Contents

General description

To the casual observer the Bully-Proof Vest appears to be a relatively harmless piece of personal body protection. It is constructed from a lightweight armoured material (composition unknown) and reinforced with riveted steel banding. This is arranged in a diamond pattern, which not only provides maximum strength but also coordinates with many popular styles of knitwear.

Mounted to the front of the garment is a small rectangular box, and it is this that houses the Bully-Proof Vest's secret weapon. When threatened, the wearer pushes a large red button on top of the box, and this releases a long, high-tension spring on the end of which is a boxing glove. The boxing glove is flattened while stored in its housing but resumes its normal size and shape the moment it is deployed. Attackers should beware as this simple device is devastatingly effective.

Once deployed, the high-tension spring and boxing glove are retrieved using a hand-operated cable winder mounted on the right-hand side of the housing. As the cable is wound in, it recompresses the spring and returns the boxing glove to its housing, ready for the next unwitting assailant.

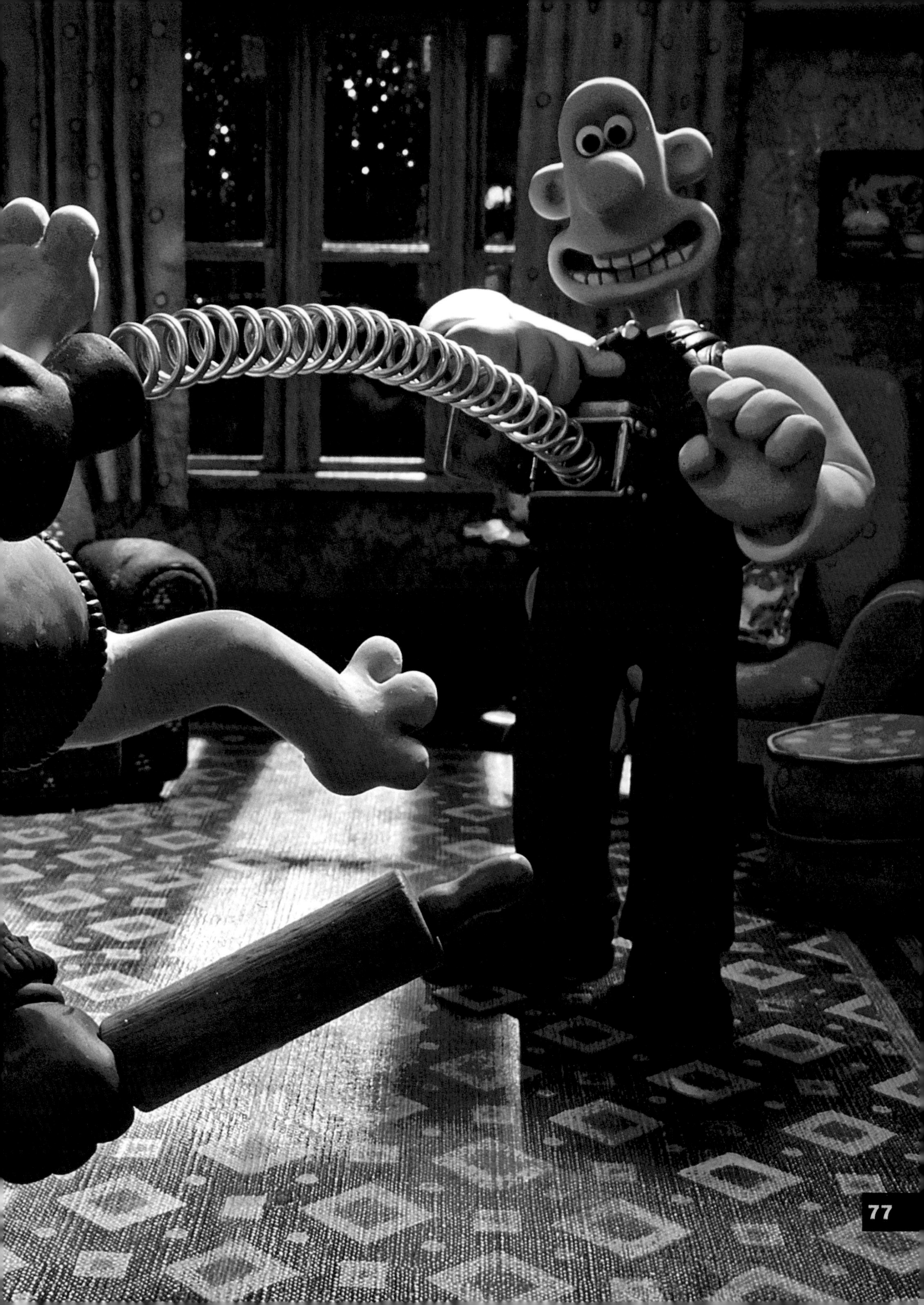

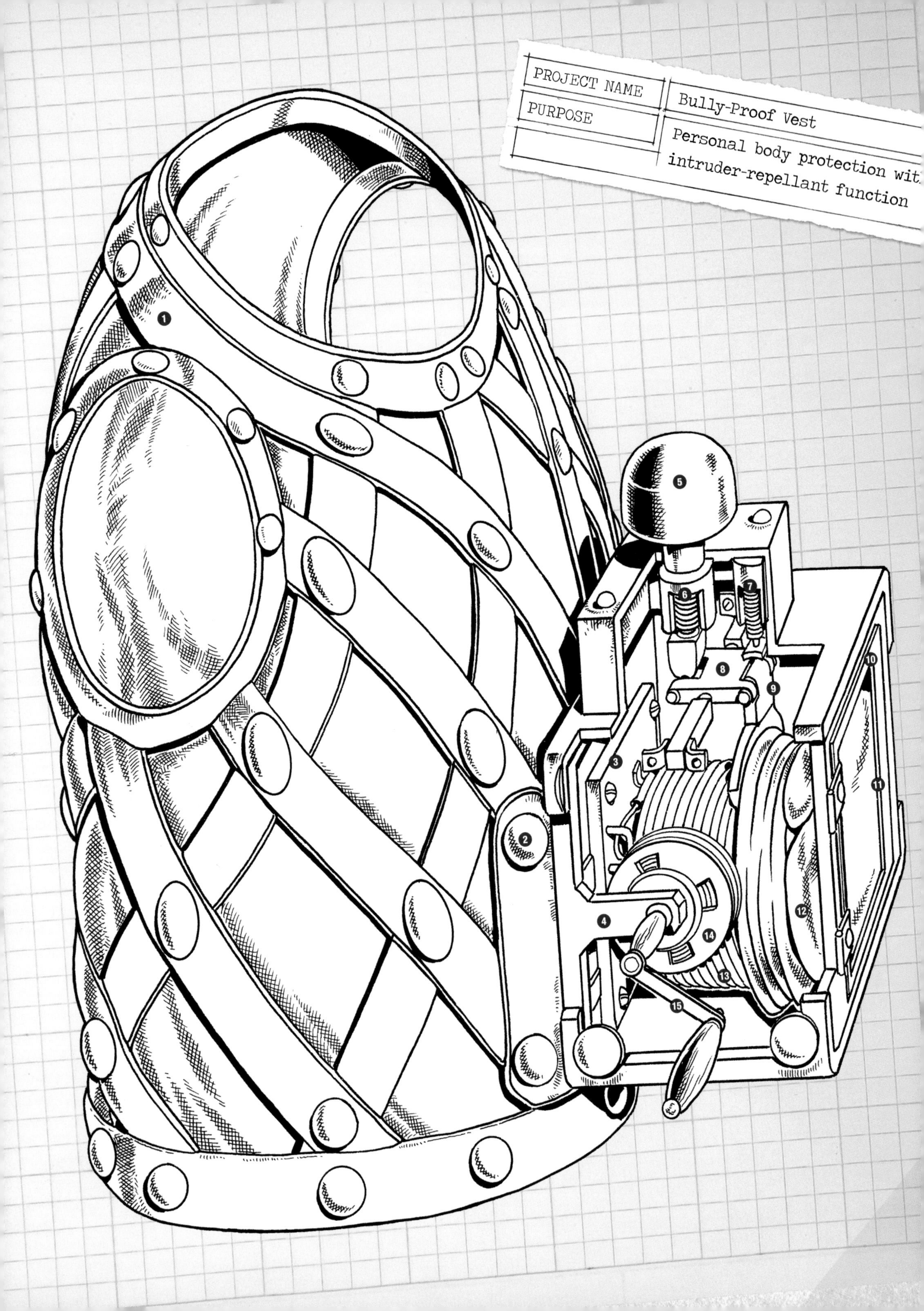
PROJECT NAME
Bully-Proof Vest
PURPOSE
Personal body protection wit
intruder-repellant function
1
2
3
4
5
6
7
8
9
10
11
12
13
14
15

Bully-Proof Vest cutaway

1. Armoured vest affords wearer protection and provides support for the boxing glove box
2. Extra strong buttons to support the weight of the boxing glove box
3. Shock-absorbing plate
4. Boxing glove deployment box
5. Boxing glove activation button (large and easy to reach in emergency situations)
6. Activation button return spring
7. Hook release cantilever
8. Cantilever repositioning spring
9. Spring release hook
10. Boxing glove box door
11. Magnetic catch keeps door shut when not in use but opens easily when boxing glove is deployed
12. Boxing glove is stored compressed behind the the box door. As it is deployed it regains its proper shape for maximum impact
13. High-tension glove deployment spring.
14. Boxing glove retrieval cable reel
15. Boxing glove retrieval winder

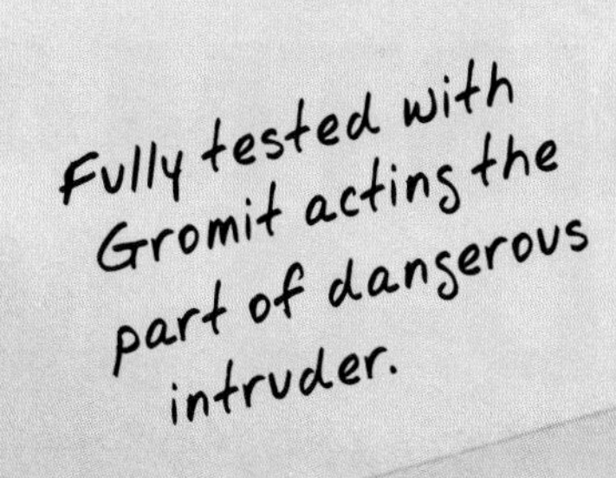

CHRISTMAS CARD-O-MATIC

Contents

I've attached a picture of our new, improved, travelling compact version

General description

Why buy Christmas cards when you can manufacture them at home? Wallace's Christmas Card-O-Matic takes a series of photographs, which are developed and then used to create a range of bespoke Christmas cards, and all from the comfort of the dining room. Our cutaway drawing shows the full-size early prototype of the Christmas Card-O-Matic, but Wallace has since developed a more compact 'portable' version. However, although scaled down the principle of operation is the same.

The camera used is of the 4x5in large-format variety and is loaded with sheet film, which is supplied from an automatic film-loading cartridge. After exposure, each film is removed from the camera and taken by the conveyor (in the dark) to the developing unit where it is developed, fixed and rinsed before being dried and passed to the printer unit. The conveyor then returns to the camera, running in a continuous loop.

Special heat-activated, self-adhesive photographic paper is used, and this is automatically loaded into and removed from the printer by an articulated grabber arm. The used films are discarded and the prints are dropped down a chute to a conveyor belt where glue and glitter are added for extra sparkle (optional).

The final stage of the process sees the printed photos applied to pre-scored cards, which are held by a vacuum holding plate, and heated to activate the adhesive backing. The finished card is then folded and placed on the exit conveyor, which carries it to a collection basket.

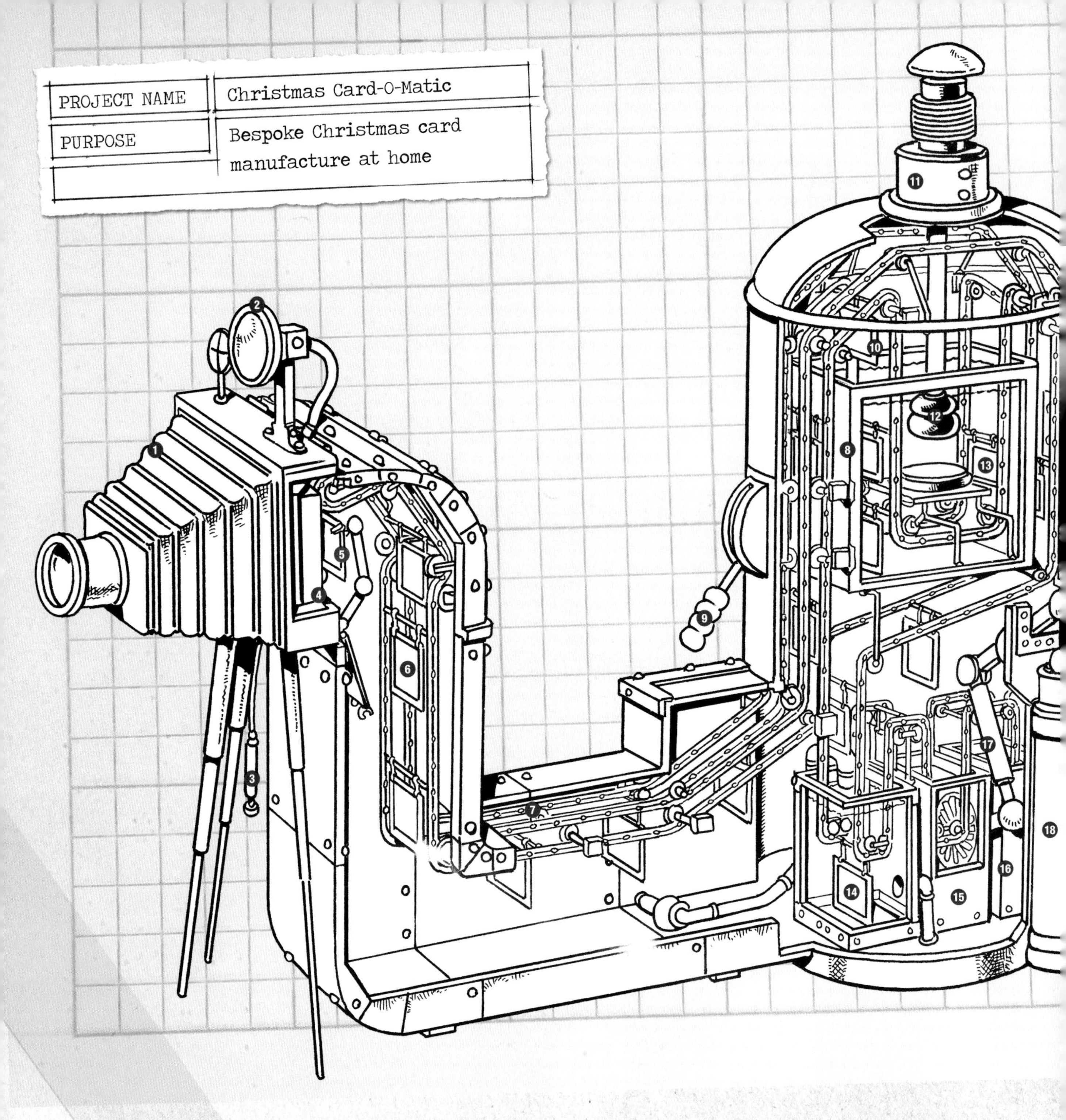

Christmas Card-O-Matic cutaway

1. Photograph is taken using 4x5in film
2. Flash gun
3. Cable release for camera
4. Automatic film-loading cartridge
5. Exposed film is removed from camera by grabbers as new film is loaded
6. In the dark, the films are hung on a conveyor and taken to the developer tank
7. Top conveyor returns to camera
8. Developer tank
9. Film activation lever
10. Films are dipped into developing tank
11. Pump circulates the developer
12. Developer circulation paddle
13. Conveyor removes films from developer tank and takes them to the water tank
14. Developer is rinsed off the films in circulated water tank
15. Films are dipped into fixer tank
16. Films are rinsed again in second water tank
17. Fixer circulation pump
18. Fixer storage tank
19. Timer
20. Drying fan

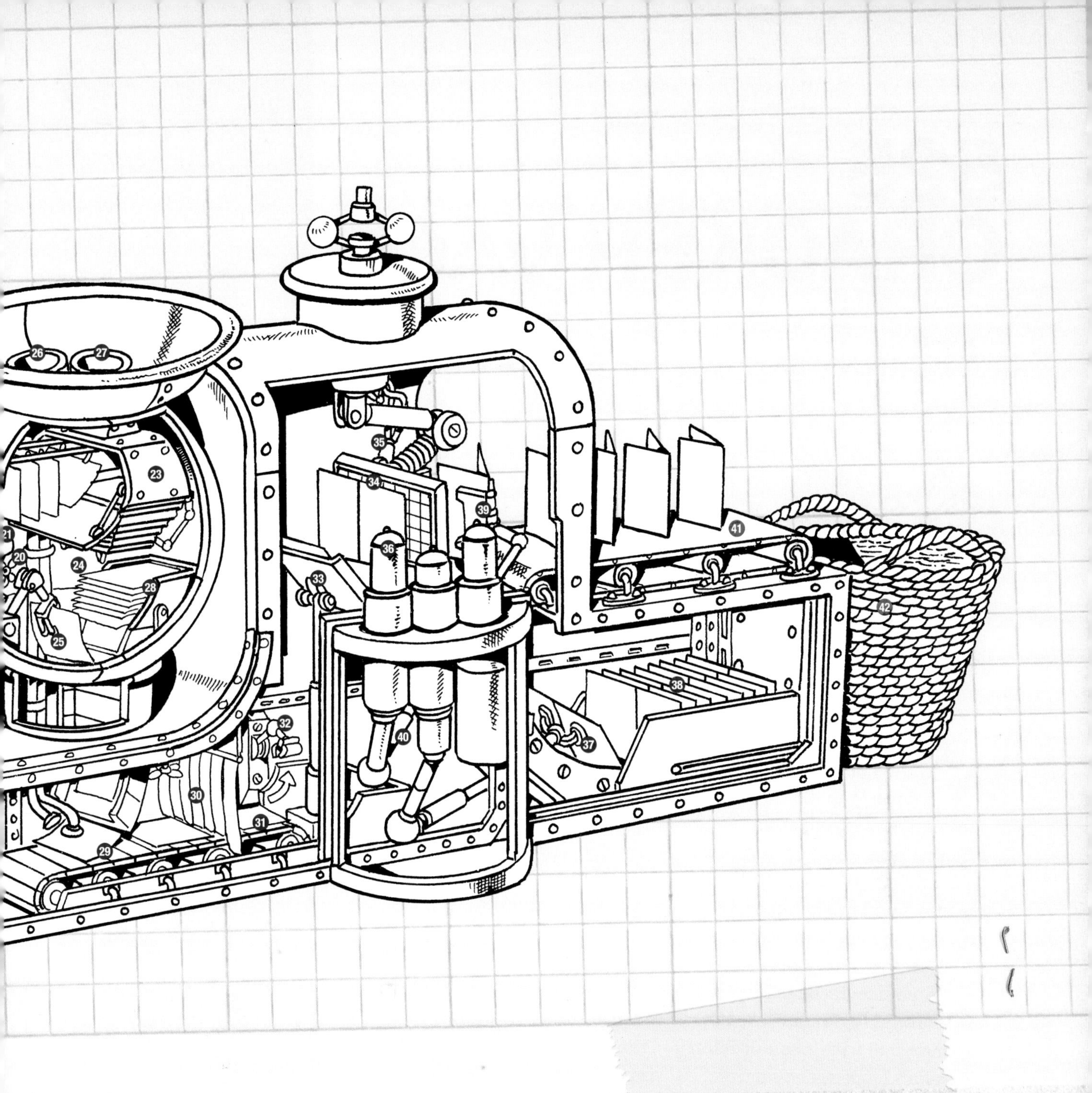

- 21 Film grabbers
- 22 Films are dried on moving rack and then placed in the printer
- 23 Film printer
- 24 Photo paper loading access panel
- 25 Grabber removes used films then places prints in chute
- 26 Glue chute used to apply glitter (optional)
- 27 Glitter chute
- 28 Pre-loaded heat-activated self-adhesive photo paper
- 29 Prints slide down chute where optional glitter is added
- 30 Light-shielding curtain
- 31 Prints are conveyed to card-assembly unit
- 32 Grabber turns print over
- 33 Prints are lifted, turned over and pressed against heated card-holding vacuum plate
- 34 Card holding plate holds card in place while photo print is glued to right-hand side
- 35 Heating conduit
- 36 Grabber hydraulics
- 37 Grabbers lift each sheet of pre-scored card up to the vacuum plate
- 38 Pre-loaded and pre-scored cards
- 39 Once photo has been glued to card, it is folded and placed on the exit conveyor
- 40 Card-folding hydraulics
- 41 Exit conveyor
- 42 Collection basket

525 CRACKER VAC

Contents

General description

Wallace describes the 525 Cracker Vac as an 'auto cleaner' with a built-in cracker sensor. It is designed to accurately locate and vacuum up stray cracker crumbs - a constant problem at 62 West Wallaby Street.

The device is based upon a traditional cylinder vacuum cleaner but with the addition of several special features, including trolley wheels with rear-wheel drive, modifications to the cleaning head and the aforementioned cracker sensor, which is a binocular visual sensing device that enables the Crackervac to locate cracker crumbs with extreme accuracy.

The 525 Cracker vac is powered by an uprated electric motor that also drives the rear wheels. This motor is powered by a rechargeable battery, allowing the cleaner complete freedom of movement around the house. The suction impeller (fan) creates a vacuum in the dust bag compartment, drawing in air (and cracker crumbs) through the cleaning head via a flexible hose. Special motor-driven teeth in the cleaning head ensure that larger pieces of cracker are broken up to prevent them getting wedged in the hose.

Unfortunately, a bug in the Cracker Vac's programming means that it is unable to distinguish between genuine dropped cracker crumbs and a half-eaten packet of crackers, a fault that is potentially hazardous to unsuspecting cracker eaters.

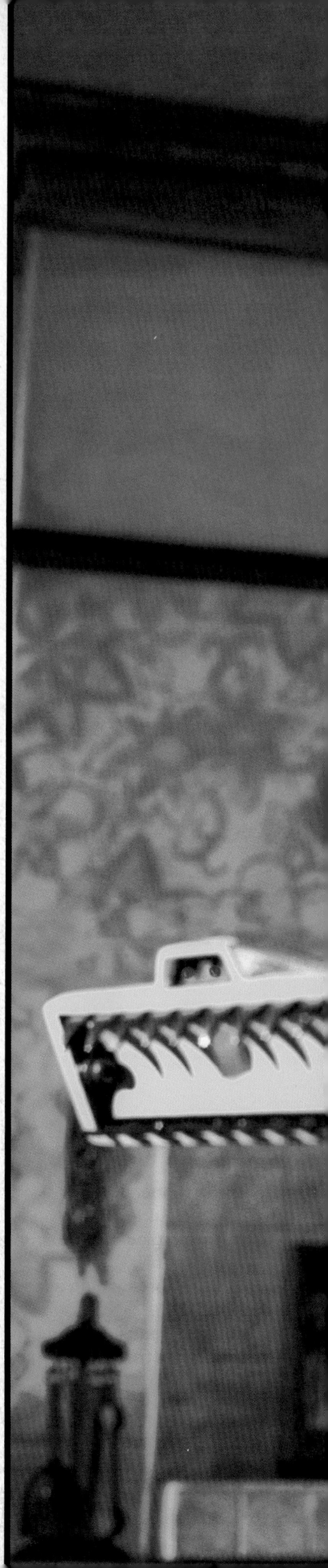

525 CRACKER VAC

I think the next version should perhaps be a little less 'automatic'

The 525 Crackervac cutaway

① Binocular cracker sensor for accurate sighting of cracker crumbs.
② Hose contains cable conduits linking cracker sensor to computer
③ Teeth grip and crush larger pieces of cracker before being vacuumed.
④ Motor and gearing for teeth
⑤ Vacuum suction duct
⑥ Dust bag securing clip
⑦ Front housing can be unscrewed to empty the dust bag when full.
⑧ Dust bag
⑨ Primary filter
⑩ Suction impeller
⑪ Secondary filter
⑫ Electric motor powers suction impeller and the rear-wheel drive
⑬ Rear-wheel gearing provides maximum freedom of movement, allowing the cleaner to vacuum in awkward places
⑭ Carrying handle
⑮ Computer and visual processing cortex enable the 525 Cracker Vac to locate cracker crumbs using data from cracker sensor
⑯ Remote-control transceiver.
⑰ Remote-control antenna, with optional duster function
⑱ Battery
⑲ Accessible Service Section (A.S.S.) can be removed to allow maintenance of motor and electronics

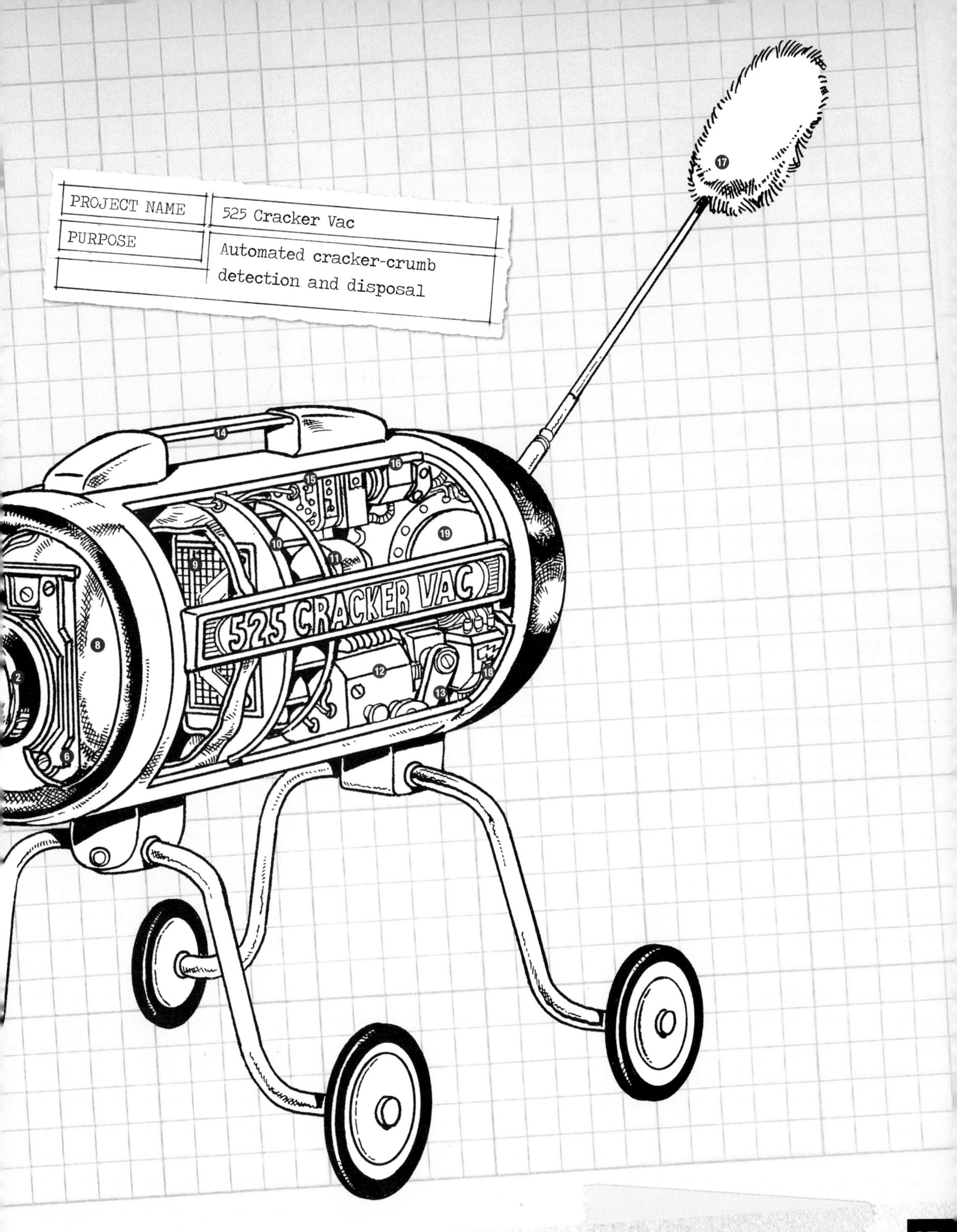
PROJECT NAME
525 Cracker Vac
PURPOSE
Automated cracker-crumb detection and disposal
525 CRACKER VAC
2
6
8
9
10
11
12
13
14
15
16
17
18
19

SHOPPER 13

Contents

It's a long loaf but I thought we might just make it

General description

Wallace's 'Shopper' is a remote-controlled, automated shopping device comprising a conventional shopping trolley to which has been added a motor driving the two rear wheels, a front wheel for steering, a video camera, two articulated arms and associated control components and wiring. The model shown here is 'Shopper 13', this being the device's 13th trip to the shops.

Shortly after the Shopper sets out on a trip (or 'mission'), compressed air expressed through nozzles is used to jettison a panel on either side of the main compartment. This allows for the deployment of two fully articulated arms and hands, which are controlled through a set of gears, pulleys and actuators on either side.

Overall navigation and command is performed by remote from 'mission control' (the cellar of 62 West Wallaby Street). On arrival at the shops, Shopper 13's mission is to locate and retrieve the 'big cheese', and this is achieved using the on-board video camera (for target identification) and the articulated arms and hands. Once safely grasped, the cheese is stowed in the main trolley compartment for the return journey.

Unfortunately, during the mission not everything goes according to plan. The cheese (a large edam) proves too heavy; the Shopper's frame starts to buckle under the load and one of the rear driving wheels falls off. The one remaining driving wheel causes the Shopper to circle helplessly in the middle of the shopping aisle. However, following some quick thinking back at mission control, a quickly extended arm grabs a nearby French stick, and uses it to stabilise the Shopper. The mission is able to continue with the Shopper using the French stick as a crutch in place of the missing wheel.

After hobbling back to West Wallaby Street, 're-entry' appears to be successful, but while scaling the doorstep to the house the Shopper becomes unstable and falls over, causing the cheese to roll out of the main trolley compartment and back down the path towards the gate.

With the edam now stranded, Wallace (as mission director) has one last option and he launches the 'probe' to try and retrieve it.

13

Shopper 13 cutaway

1. Lens
2. Iris
3. Focus lens
4. Charge-coupled device
5. Remote-control command reception antenna
6. Control systems computer and command transceiver
7. Camera and programming controls
8. Arm motor box
9. Compressed air cylinder
10. Gas jet nozzle
11. Left side panel (jettisoned after launch)
12. Right side panel (jettisoned after launch)
13. Exhaust pipe
14. Motor
15. Battery
16. Battery cover
17. Forward steering wheel
18. Left arm gearing, pulleys and control actuators
19. Trolley lid
20. Clipboard for mission number

Gromit at 'Mission Control' in the cellar of 62 West Wallaby Street

If only the 'probe' had been more reliable

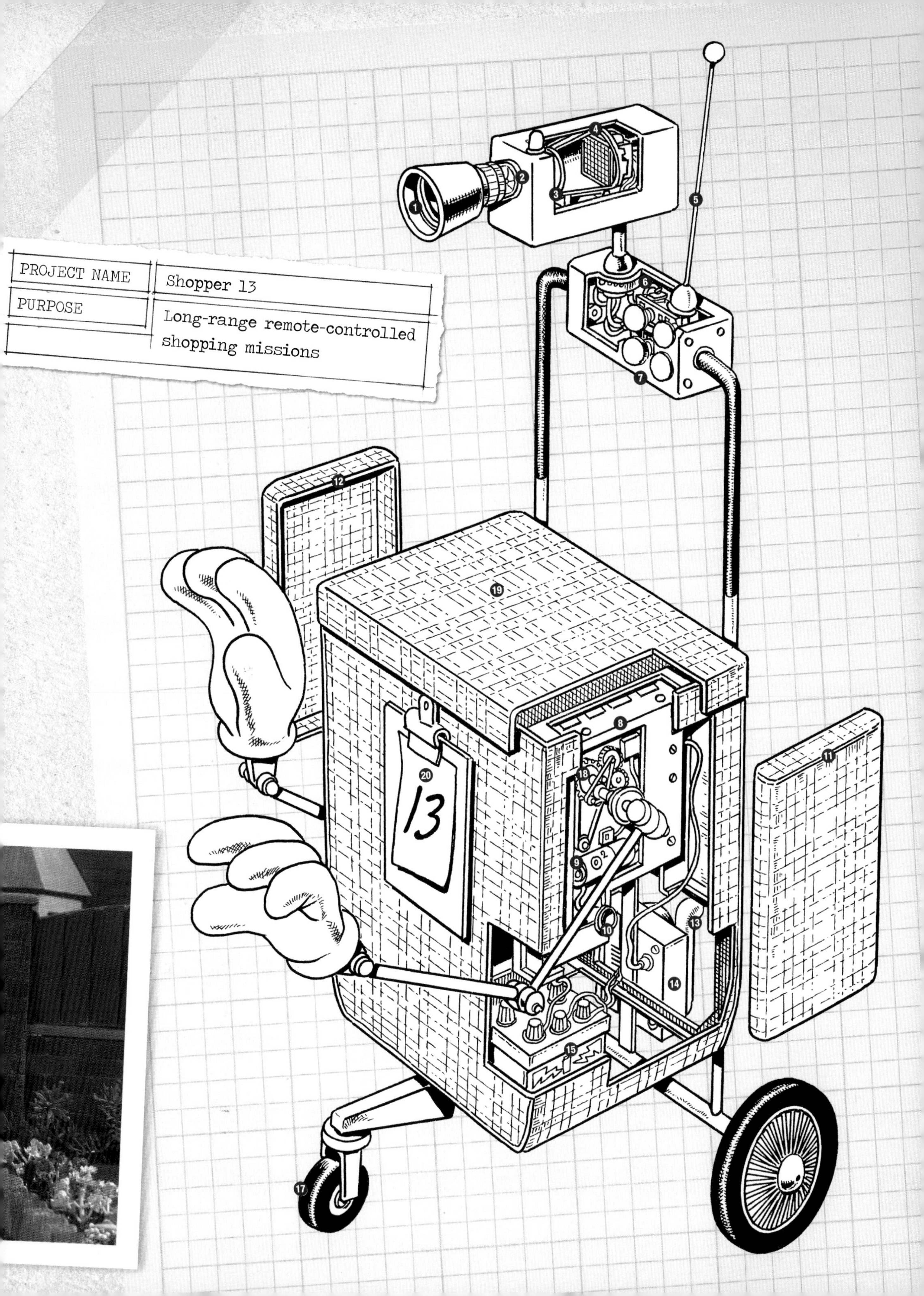
PROJECT NAME
Shopper 13
PURPOSE
Long-range remote-controlled shopping missions
13

SNOOZATRON

Contents

General description

It's three o'clock in the morning and Wallace just can't get to sleep – what can he do? Well, perhaps this is the perfect chance to try out his new Snoozatron invention: a fully featured sleep-inducing machine with guaranteed results.

Wallace activates the system by pulling a single large lever on the control box next to his bed. This sets off the Snoozatron alarm and flashing alert in Gromit's bedroom, which wakes him up so that he can go downstairs and get changed. With the exception of Gromit's role (see below), the rest of the system is completely automated and follows a pre-set program of operation.

Automatic tucking-in devices, which are mounted on articulated arms and powered by electric motors, extend through hatches in the floor on either side of Wallace's bed to tuck in the bed sheets. At the same time, motorised plumpers, mounted on hydraulic rams, descend through hatches in the ceiling, and plump Wallace's pillow from either side.

Now that the bed has been made more comfortable, a self-heating hot-water bottle and emergency teddy bear are deployed by extendable arms, again through the ceiling hatches, while a gramophone slides out on rails through a hatch in the wall to play a well-chosen selection of soothing lullabies.

With Wallace now feeling sleepy at last, Gromit is downstairs and ready for action wearing his sheep costume. From the dining room below, he is launched by the auto-return sheep-counting spring straight up and through a hatch in Wallace's bedroom floor. Wallace counts each time he sees Gromit appear at the foot of the bed, but as he eventually drifts off to sleep the counting is continued by an automatic sheep counting device attached to the top of the Snoozatron control box.

In the unlikely event that Wallace does not fall asleep then he can pull the lever to activate the Snoozatron a second time. Again the automatic tucking-in arms, pillow-plumpers and gramophone operate, and they can be activated repeatedly until sleep eventually comes. However, so far repeated activation of the system has not been necessary.

The only apparent drawback to using the Snoozatron is that Gromit must be woken up for it to operate.

0499

Snoozatron cutaway

1. Snoozatron start lever
2. Light-up auto display
3. Sheep counter
4. Cables linked to Gromit's bedroom alarm and electricity ring main
5. Primary electric motor and master control circuits
6. System wiring junction box
7. Floor of loft
8. Pillow plumping telescopic arm
9. Hot water bottle deployment arm
10. Pillow plumper control arm
11. Control arm hydraulic ram
12. Return spring
13. Hydraulic control unit
14. Padded pillow plumper
15. Electric pillow plumping motor
16. Emergency teddy deployment arm mechanism
17. Emergency teddy deployment arm
18. Emergency teddy
19. Gramophone
20. Gramophone speaker
21. Lullaby record
22. Gramophone deployment rail
23. Bedside lamp
24. Alarm clock
25. Left-hand tucking-in device
26. Tucking-in device arm control electric motor
27. Floor hatch electric motor
28. Auto-return sheep-counting spring used by Gromit when on Snoozatron duty (sheep costume not shown)
29. Self-heating hot-water bottle
30. Favourite slippers

PROJECT NAME	Snoozatron
PURPOSE	Automated, multi-function sleep-inducing system

Sheep counting continues automatically after Wallace falls asleep

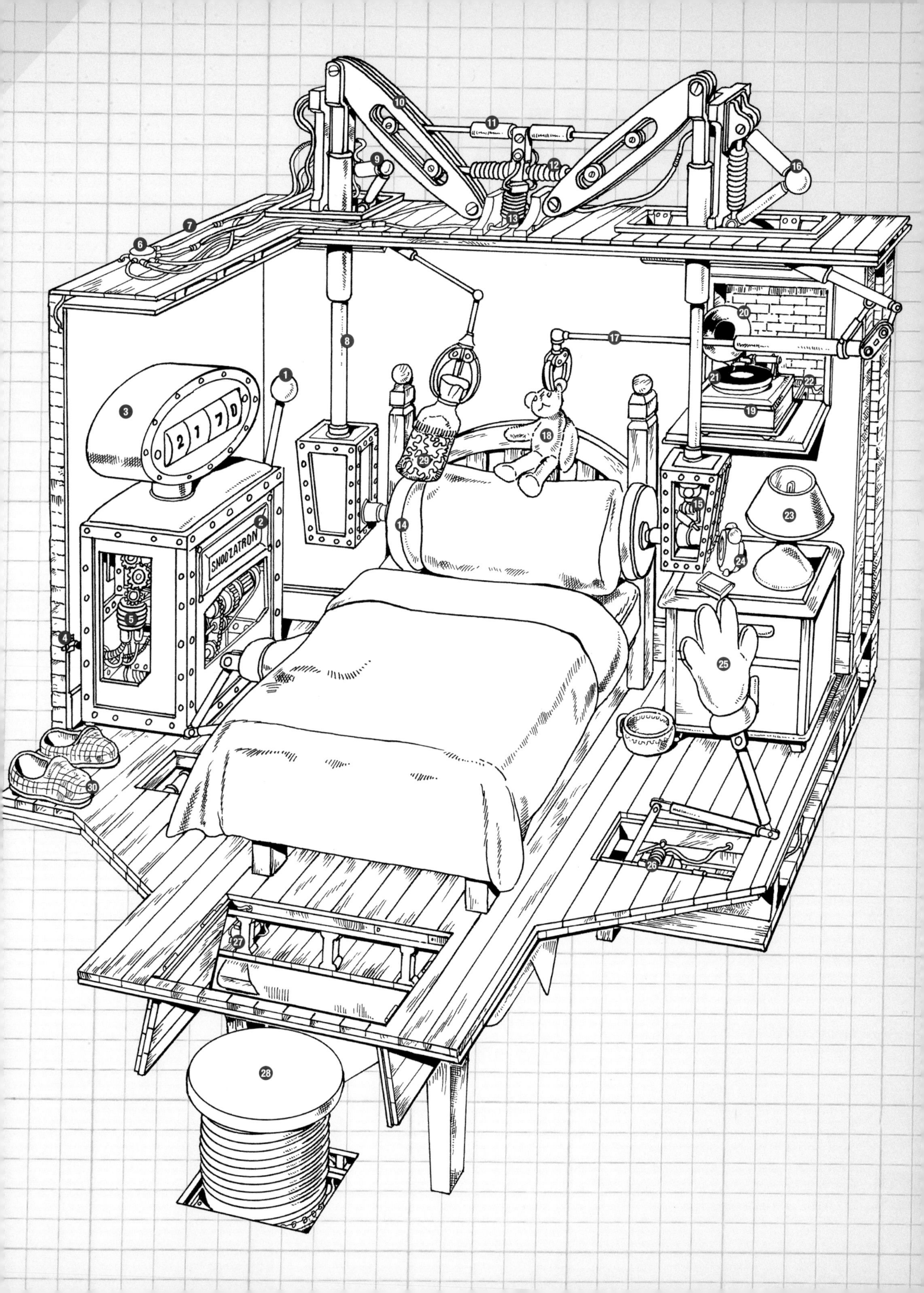
2170
SNOOZATRON

SOCCAMATIC

Contents

too much wasted space. We thought it needed more explainins

General description

When Wallace and Gromit's weekly 'beat the goalie' football shoot-outs at the local park prove to be a little too energetic (and unsuccessful) for Wallace, he invents a true labour-saving contraption in an effort to up his game.

The Soccamatic is a fully automated, motor-driven penalty-kicking machine. Once manoeuvred into position, it can launch footballs at the goal with great speed, incredible accuracy and alarming frequency.

Several footballs are stored in the primary storage tube and are ready for immediate launch via the deployment chute. A hydraulically operated pump-action mechanism ejects the balls one at a time from the bottom of the primary storage tube and down the deployment chute to four football boots mounted on a rotor. The rotor is synchronised with the pump-action mechanism, so as each ball exits the deployment chute it is kicked by one of the four boots. The rotor turns and the next ball is kicked by the next boot, and so on. The speed of delivery can be varied with no loss of accuracy, and the whole system is driven by a 400cc, two-cylinder petrol engine via a series of chains and associated gearing.

Once the balls in the primary storage tube are depleted, a secondary system can be brought into play. Five high-mounted storage cylinders rotate on a shaft, which is driven by an electric stepper motor. As the cylinders rotate around the shaft, they deliver further footballs to the primary storage tube via a capture bowl. Balls are delivered two at a time from each cylinder in turn as the mechanism rotates. This ensures consistent deployment even at high operating speeds. The height and angle of the secondary storage cylinders are adjustable and can be controlled by the operator, who also has control of ball-launch speed and trajectory as well as overall movement of the Soccamatic around the pitch.

The whole machine is mounted on four wheels, the rear two of which are driven by the engine. The front wheels are steerable, allowing the Soccamatic to be moved into the desired position.

A padded seat is provided for the operator, who can control the machine while watching the action in comfort.

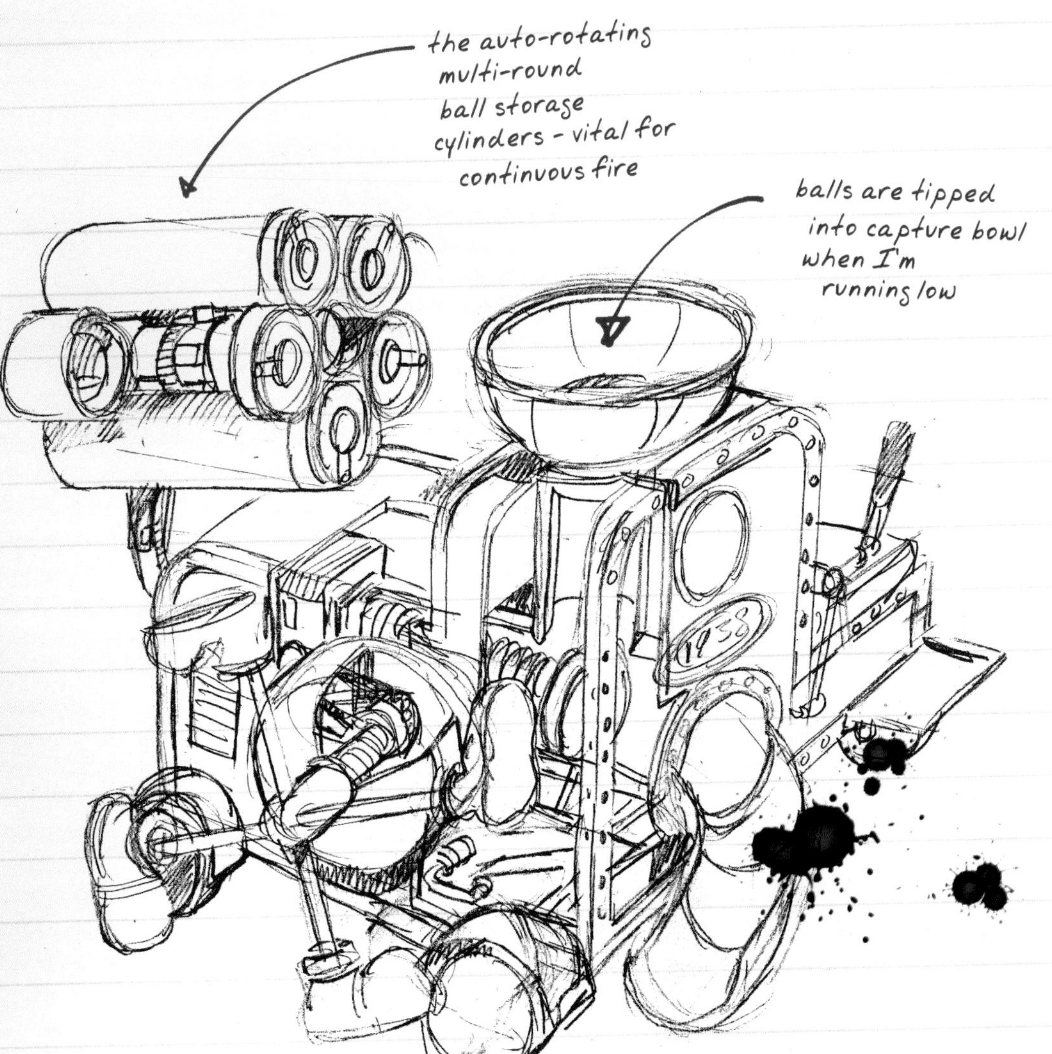
the auto-rotating multi-round ball storage cylinders - vital for continuous fire
balls are tipped into capture bowl when I'm running low
1938

The Soccamatic cutaway

1. Rotating football boot launch rotor.
2. Launch rotor gearing.
3. Launch rotor chain drive.
4. Chain drive and gear housing.
5. 400cc two-cylinder engine.
6. Fuel tank.
7. Engine cooling fan.
8. Football deployment chute.
9. Football deployment spring.
10. Football deployment pump action mechanism (synchronised to launch system).
11. Pump action mechanism hydraulics.
12. Primary football storage tube.
13. Secondary football storage cylinders.
14. Storage cylinder rotation shaft.
15. Cylinder rotation electric motor.
16. Football stop, geared to cylinder rotation to ensure balls are deployed quickly and evenly.
17. Football capture bowl.
18. Storage cylinder adjustable height support.
19. Storage cylinder height support gearing.
20. Multi-function control lever.
21. Control lever linkage.
22. Operator's seat.
23. Footplate.

PROJECT NAME	Soccamatic
PURPOSE	Motorised 'beat the goalie' football-kicking machine

13

18

16
14
15
17
20
22
21
12
23
10
11
9
8
7
19
5
2
3
4
6
1

SNOWMANOTRON

Contents

General description

It's time for the annual Grand Snowman Competition and Gromit is sculpting a fine snow statue of Wallace in 'The Thinker' pose. Meanwhile, Wallace is determined to take first prize using his new Snowmanotron invention.

As the machine is driven forward, the large front scoop fills with snow. This is then lifted up by hydraulics and loaded into a collection bowl on top of the machine behind the operator. The handbrake is applied and the machine is activated by a stop/start button on the dashboard. The articulated topside snow ram arm compacts the snow in the collection bowl and pushes it down into the snowman-making compartment, which is a heavily modified 'Smug' refrigerator. Once inside, the snow is compressed further by the topside snow ram.

Two further mechanical arms, one mounted on each side of the machine, enter the snowman-making compartment. The hands on the arms are fitted with highly articulated finger actuators, which are covered with woolly gloves to give an authentic finish.

When sculpting is complete, the operator pulls the 'eject' lever, which releases the door of the fridge and lowers the snowman deployment ramp at the rear of the machine. Finally, the completed snowman is pushed out of the fridge and down the ramp by an ejection spring and pad.

The Snowmanotron is powered by a two-cylinder four-stroke petrol engine of 648cc capacity. The engine is enclosed in a compartment located below the main snowman-making apparatus and is cooled by air forced over it by a fan, which is also driven by the engine. The air-cooled engine is light and compact if a little noisy, and drives the axle between the two rear wheels of the Snowmanotron via a chain. A second pair of wheels are positioned forward of the rear wheels but are not driven, being purely load-bearing. A single wheel at the front of the machine provides steering and additional stability.

A main battery powers the electric motor for the refrigerated snowman-making compartment, as well as the snow ram and sculpting arms, hydraulics and all electrical systems. The battery is charged by a generator, which is driven by a toothed belt from the engine.

The cockpit of the Snowmanotron is open and so operators are advised to wear suitable cold-weather clothing.

Snowmanotron cutaway

1. Snow scoop/loader bucket
2. Scoop lifting ram
3. Scoop support beam
4. Left-hand loader bucket hydraulic lift
5. Hydraulic lift ram
6. Handbrake
7. Topside snow ram start/stop button
8. Finished snowman eject lever
9. Temperature gauge
10. Speedometer
11. Dashboard electrics
12. Topside snow ram pushes snow into fridge compartment for processing
13. Finger actuators beneath outer glove ensure sculptural finesse
14. Snow compression and sculpting arms
15. Snow collection and first-stage compression chamber
16. Customised and adapted 'Smug' fridge
17. Fridge compartment keeps snow cool during sculpting
18. Expansion chamber
19. Evaporator coils
20. Refrigerant gas under pressure in condenser coils
21. Metal vanes
22. Fridge maintenance access cover
23. Fridge door handle
24. Fridge door seal
25. Flashing warning light
26. Heavy-duty temperature control curtain
27. Sculpting arm electronic control processors
28. Fridge electric motor
29. Fridge ventilation grille
30. Capillary tube
31. Engine cooling air fan
32. 650cc twin-cylinder two-stroke engine
33. Chain drive to rear axle
34. Coil ignition for each cylinder
35. Finished snowman ejection mechanism
36. Snowman ejection ramp (shown in raised position)

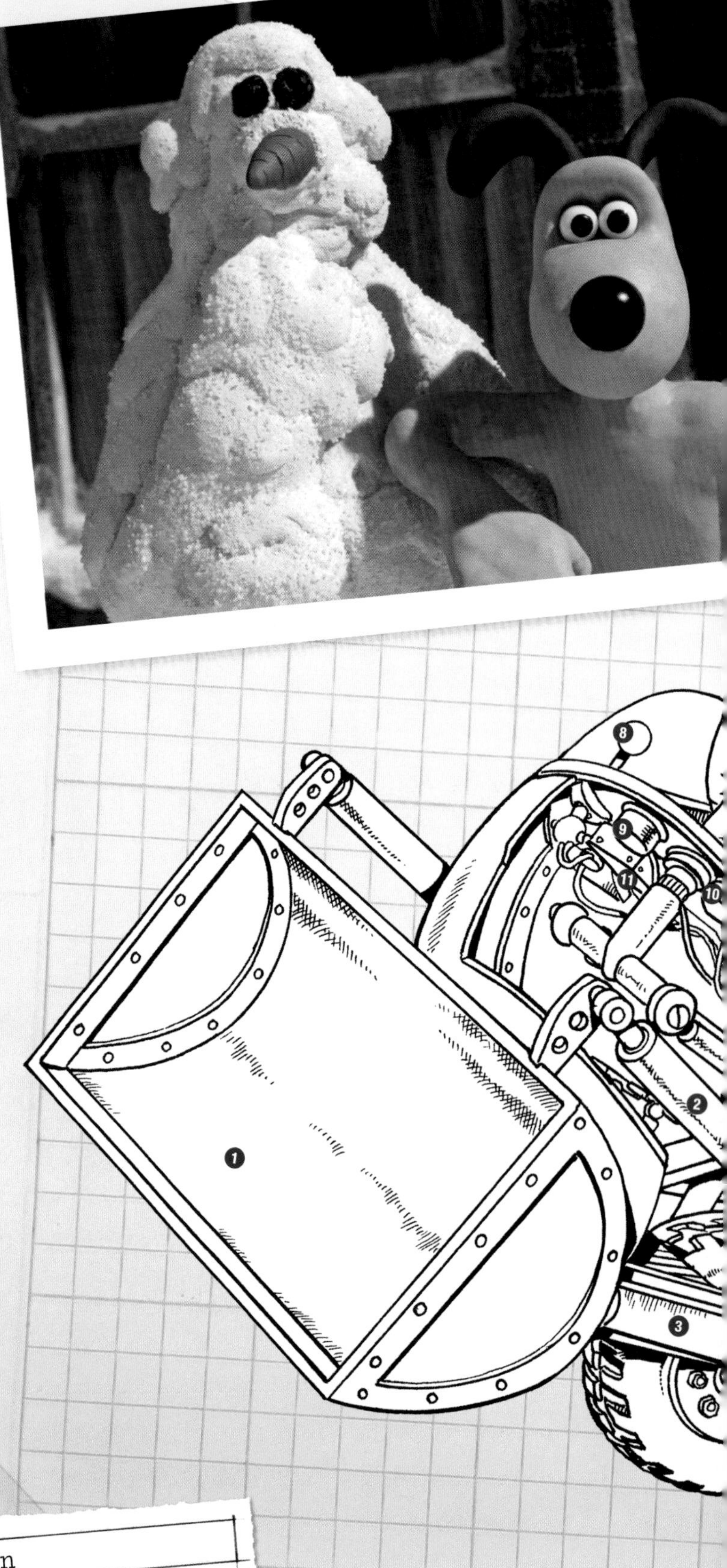

PROJECT NAME	Snowmanotron
PURPOSE	Automated competition-grade snowman-making machine

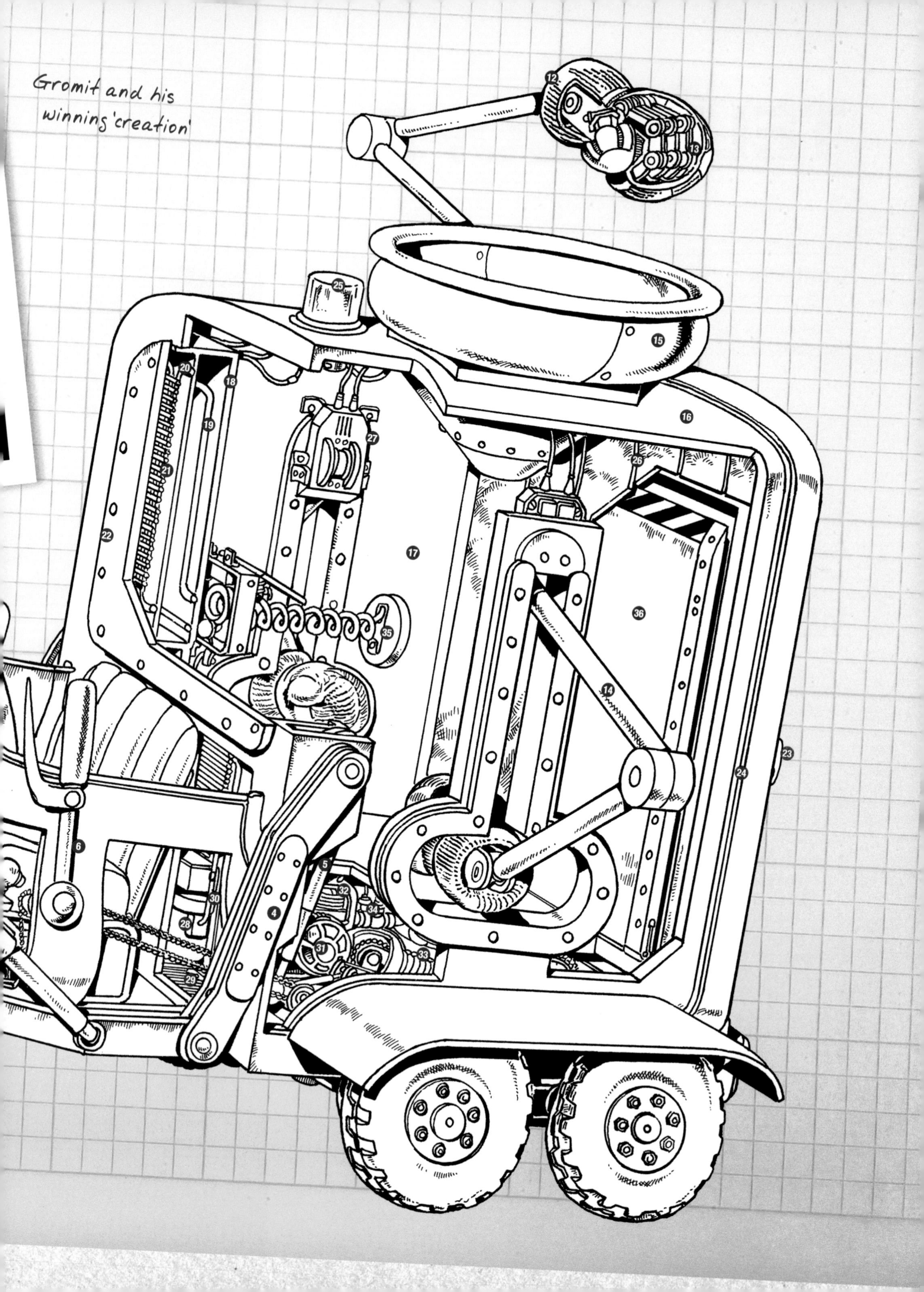
Gromit and his
winning 'creation'
12
13
15
16
17
18
19
20
21
22
23
24
25
26
27
14
35
36
6
5
4
32
34
30
28
31
33
29

TELLYSCOPE

Contents

General description

When Wallace wants to change television channels, he doesn't need to bother with the remote control when he can use his new 'Tellyscope' contraption instead.

At the touch of a button, a flap opens in the arm of the chair and up springs a gloved catapult, which Wallace loads with a tennis ball. The catapult is then pulled back and released, sending the ball flying across the lounge and through a hole in a picture on the wall above the television. Once it has passed through the hole in the wall, the ball bounces off a target pad and rolls down a straight chute, and then down a helical chute, at the end of which is another straight chute.

The ball drops from the end of the second straight chute and lands in a bucket, one of several mounted on a vertical carousel, which is turned by a motor and assisted by a counterweight. When the carousel has rotated through 180 degrees, a boot on the end of a lever kicks the ball from the bucket and into a retrieval bowl. A final chute carries the ball from the retrieval bowl down to a gloved lever, which finally activates the Tellyscope positioning beams.

The positioning beams are toothed and driven through a series of gears by an electric motor. The beams pass through holes in the wall into the lounge on the other side where they are fixed to the rear of the television, which is pushed out into the room on wheels. The otherwise-standard television is fitted with a telescopic rear casing, which shrouds the positioning beams as they extend from the wall and push the television towards the operator. An extra-long power cord ensures that the plug is not pulled from its socket as the Tellyscope extends.

With the television now within easy reach, Wallace is able to change channel or adjust the volume as required. Finally, the mechanism is reversed automatically and the Tellyscope retracts, returning the television to its usual position against the lounge wall.

The Tellyscope's only shortcoming seems to be its need for a ready supply of tennis balls, as Wallace discovers when a channel-changing emergency develops and he attempts to activate the system by catapulting the television's usual remote control instead.

MEN ARE FROM MARS
DOGS ARE FROM PLUTO

Tellyscope cutaway

1. Wallace's armchair
2. Gromit's armchair
3. Spring-loaded catapult
4. Telescopic television 'Tellyscope'
5. Extra-long power cord
6. Tellyscope positioning beams
7. Positioning beams activation level
8. Positioning beam gearing
9. Gearing for carousel and boot
10. Carousel counterweight
11. Carousel
12. Retrieval bucket

Sequence of operation

1. Tennis ball is catapulted from Wallace's armchair
2. The ball flies through the hole in the picture on the lounge wall
3. The ball bounces off the target ...
4. ... and down the chute ...
5. ... where it lands in one of the buckets; the carousel rotates, assisted by the counterweight
6. The ball is kicked from the carousel bucket ...
7. ... into the retrieval bowl ...
8. ... and down a second chute to ...
9. ... the Tellyscope positioning beam activation lever
10. The lever activates the motor and gearing which drive the positioning beams forward into the lounge
11. The Tellyscope extends into the lounge, allowing Wallace to change channel from the comfort of his armchair

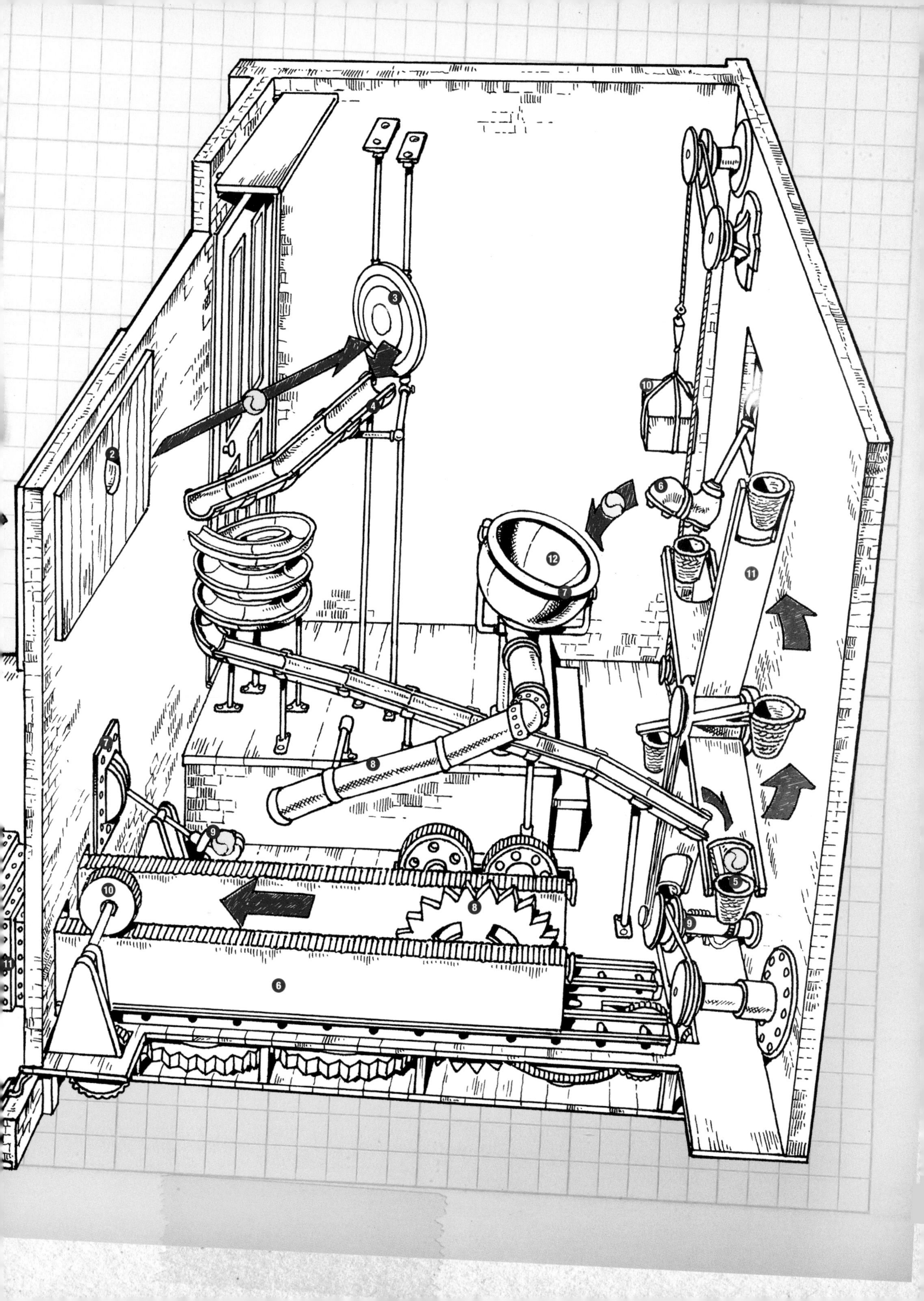

TURBO DINER

Contents

General description

The Turbo Diner is a complete, automated dining system, designed to speed up the entire process of clearing and laying the table, cooking a full dinner and clearing up afterwards.

Before activating the device, it is necessary to insert coins into a dedicated electricity meter. This activates the Turbo Diner, and it only remains for diners to take a seat and relax while the machine does all the work. However, before the system activates, arm and leg clamps fitted to the chairs secure and immobilise diners both for their own safety and also to prevent them interfering with the food preparation process.

The Clutter Clearance System consists of a large rectangular hood, which is lowered through a hatch in the ceiling to a position just above the dining table. Clutter is sucked up by a large-diameter vacuum hose,* attached to the top of the hood, and removed to the kitchen for sorting. Once the table is clear, the Clutter Clearance System retracts into the ceiling cavity, and slides to one side on rails to allow the main part of the Turbo Diner to be lowered through the ceiling hatch.

A second hood is lowered on to the dining table. This contains all the mechanisms, crockery, cutlery and ingredients necessary to prepare and serve a full roast chicken dinner (or other pre-programmed meal). Since speed is of the utmost importance, all food is cooked using an integrated, high-power microwave unit. The food is removed from the microwave and positioned on the table by a system of automated serving arms, which also lay the table with plates, cutlery and condiments.

Finally, the finishing touch to any dining occasion, a candelabra is lowered on to the table and lit automatically by a gas-powered flame-thrower (needs attention).

If there is one drawback to the Turbo Diner then it is the vast amount of power it consumes during operation.

* 300 horsepower of pure suck!

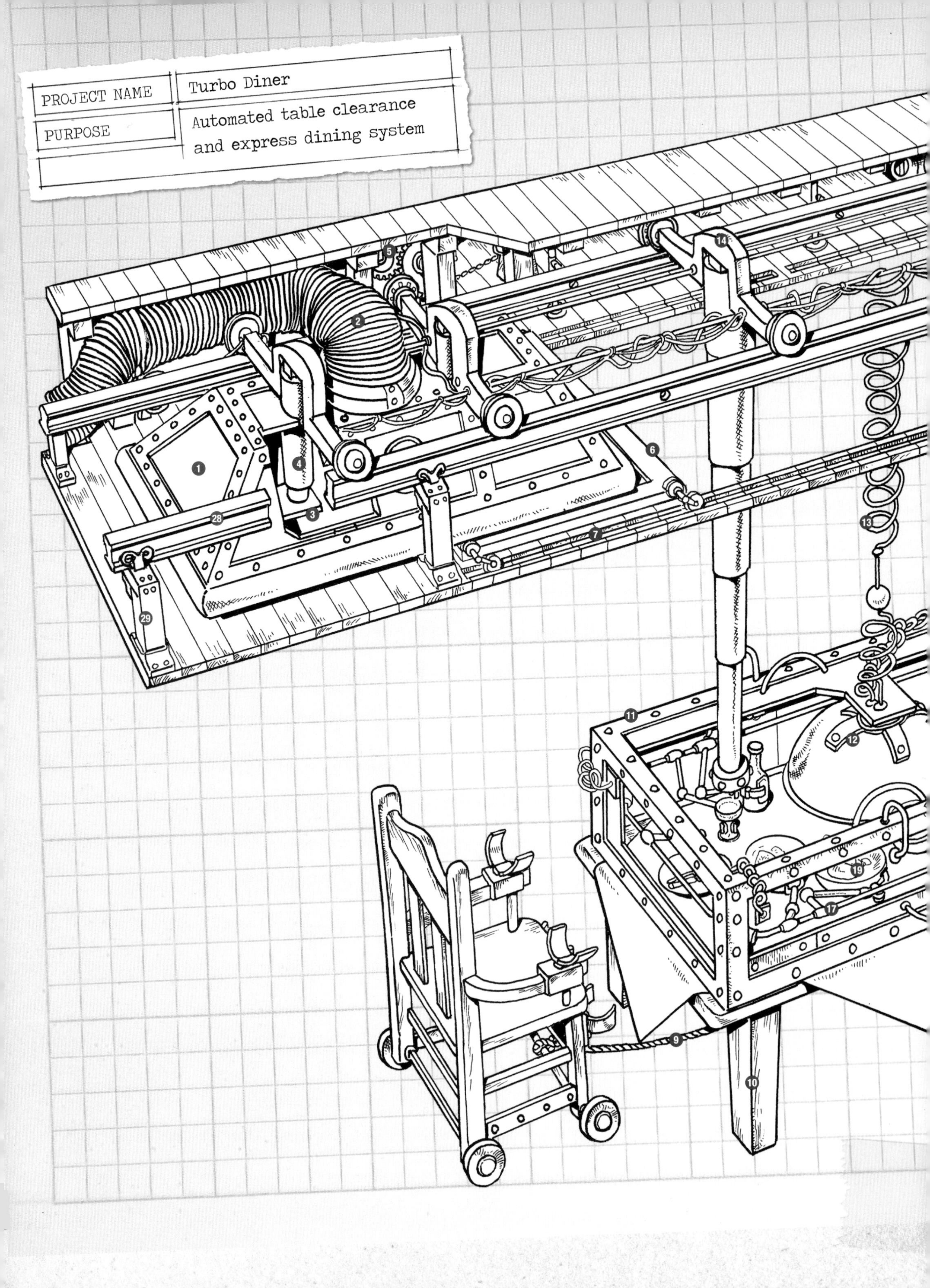
PROJECT NAME
Turbo Diner
PURPOSE
Automated table clearance and express dining system
1
2
3
4
5
6
7
9
10
11
12
13
14
17
19
28
29

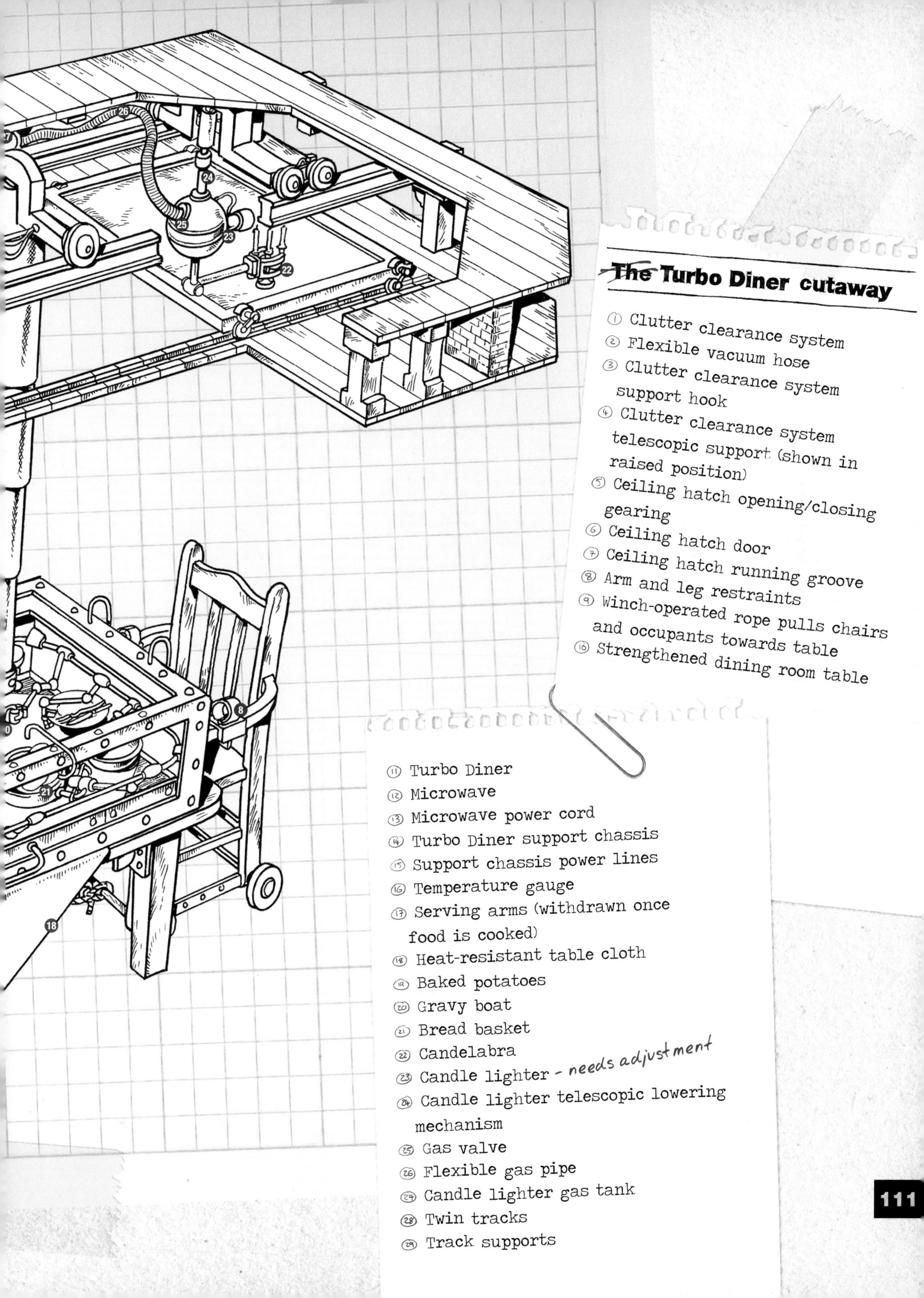
The Turbo Diner cutaway
① Clutter clearance system
② Flexible vacuum hose
③ Clutter clearance system support hook
④ Clutter clearance system telescopic support (shown in raised position)
⑤ Ceiling hatch opening/closing gearing
⑥ Ceiling hatch door
⑦ Ceiling hatch running groove
⑧ Arm and leg restraints
⑨ Winch-operated rope pulls chairs and occupants towards table
⑩ Strengthened dining room table
⑪ Turbo Diner
⑫ Microwave
⑬ Microwave power cord
⑭ Turbo Diner support chassis
⑮ Support chassis power lines
⑯ Temperature gauge
⑰ Serving arms (withdrawn once food is cooked)
⑱ Heat-resistant table cloth
⑲ Baked potatoes
⑳ Gravy boat
㉑ Bread basket
㉒ Candelabra
㉓ Candle lighter - needs adjustment
㉔ Candle lighter telescopic lowering mechanism
㉕ Gas valve
㉖ Flexible gas pipe
㉗ Candle lighter gas tank
㉘ Twin tracks
㉙ Track supports

"Right everyone, now it's your turn to be inventors. Get your thinkings caps on!"

ACKNOWLEDGEMENTS

The author and Haynes Publishing would like to thank the following people for everything they have contributed to the production of this book:

Graham Bleathman for his incredible cutaway drawings, attention to detail and sheer dedication, without which a book like this would not be possible;

Lee Parsons for the patience and imagination he put into designing the pages and cover; and

everyone at Aardman Animations, and particularly Neil Warwick for responding to umpteen image requests, Merlin Crossingham for taking time out to look at artwork and showing genuine enthusiasm throughout, Dick Hansom for brilliant additional material and Tresta Baber for invaluable support and helping to make the whole thing happen.

Finally, a HUGE thank you to Wallace and Gromit for allowing access to all the marvellous contraptions featured in this book and for providing additional insights into their design and inner workings.

PICTURE CREDITS

Cutaway drawings and Edit-O-Matic by Graham Bleathman.
All other images courtesy of Aardman Animations.